OIL

OIL

The present situation and future prospects

ORGANISATION FOR ECONOMIC CO-OPERATION AND DEVELOPMENT   PARIS 1973

The Organisation for Economic Co-operation and Development (OECD) was set up under a Convention signed in Paris on 14th December, 1960, which provides that the OECD shall promote policies designed :

— *to achieve the highest sustainable economic growth and employment and a rising standard of living in Member countries, while maintaining financial stability, and thus to contribute to the development of the world economy;*

— *to contribute to sound economic expansion in Member as well as non-member countries in the process of economic development;*

— *to contribute to the expansion of world trade on a multilateral, non-discriminatory basis in accordance with international obligations.*

The Members of OECD are Australia, Austria, Belgium, Canada, Denmark, Finland, France, the Federal Republic of Germany, Greece, Iceland, Ireland, Italy, Japan, Luxembourg, the Netherlands, Norway, Portugal, Spain, Sweden, Switzerland, Turkey, the United Kingdom and the United States.

*
**

CONTENTS

FOREWORD 11

SUMMARY AND CONCLUSIONS 13

Chapter I

OIL, ENERGY AND THE OECD 19

Introduction 19
A. Oil in OECD Europe's economy 21
B. Oil in the North American economy 25
C. Oil in Japan's economy 28
D. Oil in Australia's economy 30
E. Oil and the rest of the World 32
Summary Oil and World energy demand 33

Chapter II

THE AVAILABILITY OF OIL 51
A. Today's proved world oil reserves 51
B. Obtaining reserves in the future 52

Chapter III

SOURCES OF OIL SUPPLIES 57
A. World crude oil production 1965-1970 57
B. The sources of OECD Europe's oil supplies 58
C. The sources of North America's oil supplies 61
D. The sources of Japan's oil supplies 64
E. The sources of Australia's oil supplies 65

Chapter IV

THE SUPPLY OF OIL 71
A. The state of the oil market 71
B. The increasing dependence of the OECD area on imported oil supplies
   and the reliability of these supplies 75
   a) The increasing dependence 75
   b) The reliability of supplies 77
C. Developments in government policies on oil and oil supply 78
The oil producing countries 78
The oil consuming countries 82

Annex

Examples of national oil company-contractor arrangements 91
a) Joint venture arrangements 91
b) Production-sharing contracts 91
c) ERAP-type contracts 92
d) Venezuelan-type service contracts 92

Chapter V

THE REFINING INDUSTRY 93
A. General introduction 93
B. Refinery development in Europe 95
C. Refinery development in North America 98
D. Refinery development in Japan 100
E. Refinery development in Australia 101
F. Future trends 102

Chapter VI

TRANSPORT 109
A. Sea transport 109
B. Pipelines 114
C. Harbour facilities 117

Chapter VII

ENVIRONMENTAL PROBLEMS 137
A. Introduction 137
B. The petroleum industry 137
     Production 137
     Refineries 137
     Atmospheric pollution by refineries 140
     Transport 141
     Storage 145
C. Air pollution by stationary installations 145
D. Air pollution by non-stationary installations 147

Chapter VIII

INVESTMENT REQUIREMENTS FOR A CONTINUING DEVELOPMENT 153
A. The background and fundamental aspects of oil investment 153
B. Investment in the United States and Canada 154
C. Investment in the OECD European area 154
D. Investment in the Middle East, Africa, Venezuela and the Far East 154
E. Future investment requirements and how they can be met 155

Chapter IX

NATURAL GAS 161

Chapter X

OIL PRODUCT PRICES 163
A. Trends during 1960-1969 163
B. Developments in 1970 and 1971 164
C. Future trends 165

ANNEXES

I. Country chapters 177
II. Statistical tables 259

TABLES

Chapter I

1. World's primary energy requirements, 1960-1980 41
2. OECD Europe's primary energy requirements, 1960-1980 42
3. OECD Europe's oil products consumption pattern, 1960-1980 43
4. North America's primary energy requirements, 1960-1980 44

5. North America's oil products consumption pattern, 1961-1970 45
6. Japan's primary energy requirements, 1960-1980 46
7. Japan's oil products consumption pattern, 1960-1970 47
8. Rest of the world's primary energy requirements, 1960-1980 48

Chapter II

9. Proved world oil reserves at end 1971 55

Chapter III

10. World crude oil production 1965 and 1970 67
11. OECD Europe crude oil supply 1960, 1965 and 1970 68
12. OECD North America oil supply 1960, 1965 and 1970 68
13. Japan's oil supply 1955, 1960, 1965 and 1970 69
14. Australia's oil supply 1960, 1965 and 1970 70

Chapter IV

15. Libyan agreement on posted prices and taxes, Mediterranean, 1970-1975   85
16. Iranian agreement on posted prices and taxes, Persian Gulf Shipments,
    1970-1975 86
17. Saudi Arabian agreement on posted prices and taxes, Persian Gulf ship-
    ments, 1970-1975 87
18. Iraqi agreement on posted prices and taxes at Eastern Mediterranean (via
    IPC pipelines), 1970-1975 88
19. Saudi Arabian agreement on posted prices and taxes at Eastern Mediterra-
    nean (via Tapline), 1970-1975 89

Chapter V

20. Refinery capacity in the Caribbean and the Middle East, 1960, 1965 and
    1970 104
21. Refinery capacity in OECD Member countries, 1960, 1965, 1970 and 1975   105
22. Trends in refinery output, 1960-1970 106
23. OECD Europe refinery location 107
24. Average intake capacities of refineries, 1965 and 1970 108

Chapter VI

25. World tanker fleet at end 1971 (by age, size and propulsion) 127
26. World tanker fleet (by size) for years 1960-1971 128
27. World scale nominal tariff (as of 1st January, 1972) 129
28. World tanker fleet at end 1971 (by flag and ownership) 130
29. World tanker fleet at end 1971 (by size and ownership) and new buildings
    on order (by size and ownership) 131
30. Some important pipelines: crude oil, petroleum products, and natural gas   132
31. Harbours in OECD countries accessible for tankers of 200,000 DWT and
    larger and possible expansion plans 134

Chapter VII

32. Weighted average loads for main contaminants in refinery effluents in
    Western Europe (in kilograms per 1,000 tons of crude oil intake) 150
33. Weighted average of waste loads in Europe and the United States, as a
    function of effluent treatment (in kilograms per 1,000 tons of crude oil
    intake) 151
34. Quantities of liquid fuel used in stationary installations in OECD countries
    in 1969 152

Chapter VIII

35. World capital expenditures in the domain of oil for the period 1960-1970   155
36. Comparison of world shares in capital expenditure in the domain of oil
    for the 1953-1962 decade with the 1960-1970 decade 159
37. World investments in the domain of oil in gross fixed assets— at end 1970   159
38. Capital dollars spent in 1970 in production per barrel of oil produced ..   160

Chapter X

39. Breakdown of an average price of a barrel and a ton of crude oil as paid
    by the consumer in selected West-European countries 176

FIGURES

1. Primary energy requirements in OECD Europe (1960 to 1980) 35
2. Primary energy requirements in North America (1960 to 1980) 36
3. Primary energy requirements in Japan (1960 to 1980) 37
4. Primary energy requirements in the rest of the world (1960 to 1980) 38
5. World's primary energy requirements (1960 to 1980) 39
6. Primary energy requirements by area (1960 to 1980) 40
7. Tanker voyage charter rates and average freight rate assessment (AFRA) 123
8. Comparative transportation costs, by various sizes of tankers 124
9. Transportation cost for a 16-inch diameter pipeline 125
10. Typical pipeline cost curves 126
11. Development of prices for gas/diesel oil at selected cities of the European Community 167
12. Development of prices for heavy fuel oil at selected cities of the European Community 168
13. Development of prices for motor gasoline (premium) in several countries of the European Community 169
14. Development of prices for regular grade gasoline at selected cities in the United States 1960-1970 170
15. Development of average refinery prices for No. 6 fuel oil at selected cities in the United States, 1960-1970 171
16. Development of prices for gas/diesel oil in Japan 172
17. Development of prices for heavy fuel oil in Japan 173
18. Development of prices for motor gasoline in Japan 174
19. How the price of a gallon of gasoline is made up in the United States .. 175

MAPS

1. International flow of petroleum 1967 (before the Arab-Israeli war) 121
2. International flow of petroleum 1970 122

FOREWORD

It is now eight years since the OECD Oil Committee published a report reviewing the main features of oil supply and consumption in the OECD area. The present report covers the past decade and indicates the probable trends to 1980. Reflecting the increase in the importance and complexity of the subject, it is considerably more comprehensive than the last, even though it does not now deal at length with natural gas, on which the Oil Committee reported separately recently. In particular, the report now has extensive sections concerned with the development of Government policies in the field and with environmental aspects.

In forecasting the probable pattern in 1980, there is no pretence of prophetic infallibility. The present supply and consumption situation and outlook result from political and other changes affecting the oil industry all over the world. It is clear that the position in 1980 will also result from a wide range of changes, some of which cannot now be foreseen. Indeed, this report was completed in mid-1972, and inevitably some issues, notably those arising in the negotiations between the producing country governments and the international companies will have moved on. However, taking the report as the latest of a series concerned with supply and demand questions over the past two decades it must be observed that, despite rapid and indeed violent changes in some factors, the oil position nevertheless displays considerable overall stability. In the outcome, change in the oil industry is massive rather than catastrophic.

In the OECD area, and indeed throughout the world, oil continues to grow both relatively and absolutely as a source of energy and of raw materials. The problems involved in determining policies become steadily more intricate; it is to be hoped that this report will be of interest and assistance to all those who are concerned one way or another with the complexities of oil policy, and in particular that it will help those who have to explain such policies to the public.

J. Angus BECKETT
Chairman, Oil Committee

June 1972.

SUMMARY AND CONCLUSIONS

1.   During the 1960s the world demand for primary energy rose by 4.9 per cent per year.   The relative position of solid fuels continued to decline, following the pattern established in previous years, while the consumption of liquid fuels accounted for well over half the absolute growth in energy utilization.   Far behind oil in world significance but exhibiting the highest growth rate over the past decade is natural gas.

2.   It is anticipated that the world demand for energy during the present decade, buoyed by accelerating requirements in the developing countries, will expand at an average annual rate— 5.6 per cent— exceeding that experienced between 1960 and 1970.   The position of oil will be further solidified and its contribution will exceed that of solid and gaseous fuels combined.   The most rapid growth among the individual forms of energy will be registered by nuclear power, especially in North America, stimulated by conventional fuels supply restrictions and by environmental consideration.

3.   Oil, with its price advantages, convenience in transport and use, and above all its plentiful supply, emerged as OECD Europe's major source of energy by the mid-1960s.   But absent indigenous sources, virtually all of the growth in the use of oil had to be met through imports, with the result that by 1970 almost 60 per cent of all OECD European primary energy requirements were satisfied by purchases from abroad.

4.   Looking ahead to 1980 for OECD Europe, oil— and gas— are likely to retain their dominance of the European energy market, as per capita demand for energy may increase by a further 50 per cent.   An important portion of the projected demand for oil and gas will be captured by indigenous sources as North Sea oil and gas come into their own.   Enlarged demands for coking coal will not be sufficient to offset further declines in the relative contribution of coal to total energy supplies, although it should be emphasized that supply and demand for all fuels, and especially for coal, may be manipulated by government policy.   In the current decade nuclear power is expected to show substantial development, supplying perhaps 7 per cent of total energy requirements by 1980.

5.   The development of energy demand in North America has been quite different than that in Europe. Neither coal nor oil were able to keep pace in relative terms with the expansion in total energy demand of 4.5 per cent per year.   Their loss in importance may be attributed to the presence of natural gas available in plentiful supply and offered at very convenient

prices, which was able to improve its market position at the expense of coal and oil.

6.   Energy in North America during the past decade differed in another important aspect, in that during the latter half the demand for energy increased more rapidly than did the growth in Gross National Product. Normally for industrially developed nations or areas, the growth in energy utilization lags behind the growth in GNP. It is thought that this apparent deviation was brought about by the greater use of electricity in homes and industry and by a decrease in opportunities to improve efficiencies in energy use.

7.   Further, North America has been able to cover the bulk of its oil requirements through indigenous production. Yet a part of this advantage is to be lost in coming years as restrictions on domestic oil production imposed by cost, environmental or other considerations will enlarge the gap between demand and indigenous supply. As a result, it is thought that by 1980 North America will look abroad for nearly 35 per cent of the supplies required to cover demand.

8.   Moreover, energy demand is projected to increase during 1971-80 at an average annual rate of 5 per cent, exceeding the growth rate of the previous ten-year period. Among the various forms of energy, oil is expected to maintain its present share of the market, coal will continue to decline and a drop in the natural gas share is also predicted. The declines in the relative importance of coal and natural gas may be attributed to the emergence of nuclear power in the 1970s. Indeed, the projected growth in nuclear power, if realized, will exceed the growth in demand for oil.

9.   Energy consumption in Japan grew at just under 12 per cent per year during the past decade, with the overwhelming portion of this growth supplied by oil. In turn, virtually all of the oil has to be imported in view of the absence of indigenous sources.

10.   Despite an anticipated decline in energy and oil growth rates, reflecting a concern for the environment and a more rational use of energy, by the end of the present decade Japan will be more dependent upon oil supplies and in turn on imports, than any other OECD Member country.

11.   Difficulties associated with ensuring a continuing and stable supply of oil, at prices acceptable to both producer and consumer, are not expected to stay the growth in demand for this fuel. Oil demand growth is forecast to exceed the average annual growth for all forms of primary energy and individually for solid and gaseous fuels as well. In sum, we forecast a demand world-wide for oil reaching some 4 billion tons (the equivalent of 80 million barrels per day) by 1980. Of this world-wide demand, as much as 70 per cent may originate in the OECD area.

RESERVES

12.   World proved reserves of crude oil are more than ample to satisfy projected needs for the foreseeable future. The bulk— almost two-thirds— of world oil reserves are to be found in the Middle East and it is unlikely

14

that the importance of this area, either as a producer or as a holder of reserves, will be diluted during this decade. But while we are confident that ample potential exists for sustaining the current increasing rate of production for many years ahead, environmental considerations and concern for the conservation of oil resources may preclude the realization of this potential at a pace consonant with past trends.

SOURCES OF OIL SUPPLIES

13.   Since the publication of the 1964 report "Oil Today", a number of important shifts in world crude oil production have taken place. First, North America has lost its place as the largest single oil-producing area to the Middle East, with the latter now accounting for some 29 per cent of the world total. Second, Africa has come to the fore as an important supplier of oil, having benefitted greatly from its geographical proximity to markets in Europe. Of all of the major producing countries, the United States exhibited the lowest rate of growth during the 1960s, illustrating again the difficulties of expanding upon a crude oil producing base of advancing maturity. The Communist Countries were not able to improve on their relative position, despite a substantial absolute growth, and accounted for one-sixth of the world total in 1970.

14.   OECD Europe over the past ten years has been able to reduce its dependence upon the Middle East as a source of oil supply, from 73 per cent to 50 per cent. In parallel the relative share of imports from Africa have grown from just 5 per cent in 1960 to more than one-third in 1970. It is thought that in the years ahead the Middle East may be able to regain a part of its lost position as an oil supplier to Europe.

15.   The search for and discovery of oil and gas in the North Sea has dominated exploration activities in OECD Europe. While precise forecasts of future production levels from these finds are not feasible, it is thought that by 1980 an output of 150 million tons might be approached, which would be able to cover 15 per cent of European oil requirements.

16.   Eastern Europe, for a variety of reasons, political and economic, has begun to import oil and such imports are likely to grow faster in the years ahead than will exports. Thus a weakening of the net trade position of Eastern Europe can be expected.

17.   For North America as a whole the development of the Arctic and of the off-shore areas offers great promise for new supplies. Yet access to these areas and the ultimate development of production will have to be responsive to a great variety of environmental considerations. The potential for oil and gas in the United States exceeds by far the performance of the domestic industry.

18.   Delays in developing new supplies and the inability of the older established fields to respond to higher demands mean that for the United States imported oil will be looked upon as the principal supplementary source of supply for the remainder of this decade.

19.   In 1970 Japan imported a total of nearly 170 million tons of oil to satisfy its burgeoning demand. Of this amount, 85 per cent originated

in the Middle East and most of the remainder came from the Far East, largely Indonesia. Other sources have never participated to any extent in the Japanese oil market, with the exception of Eastern Europe which at one time supplied 5 per cent of Japanese total crude oil imports.

THE SUPPLY OF OIL

20.   The state of the oil market as it entered the decade of the 1970s was completely unlike that which had existed at the beginning of the 1960s. Troublesome oil surpluses had vanished, the buyer's market had been transformed into a seller's market and the host nations had been able to secure for themselves a greater control over the market. All this brought inevitable upward changes in consumer prices.

21.   The demand for oil in the OECD Member countries is projected to increase from 1,573 million tons in 1970 to 2,788 million tons by 1980. Expansion in indigenous supplies will be sufficient to cover little more than 30 per cent of this indicated growth. Thus the emerging pattern is one of increased dependence on imports which in turn implies that energy policies and related political actions and programmes will have to take into account the implications of this presumed enlarged dependence on foreign supplies.

22.   Increasing purchases of oil abroad point to the importance of the reliability of these supplies. Supplies can be interrupted for technical, commercial or political reasons. Present circumstances in the world oil market underscore that the danger of interruptions of normal supplies will be greater in the years to come than in the past. The need to seek all possible measures of stabilisation is therefore of increasing importance. As one move, with this in mind, the consuming countries are taking steps to improve stock levels to the recommended 90-day level.

23.   In the latter years of the 1960s and especially at the beginning of the 1970s the pace of change in relationships between the international oil companies and the host nations accelerated markedly. The Organization of Petroleum Exporting Countries, drawing upon its newly-found cohesiveness, has been able to secure major advances in posted prices and in their oil-derived revenues, national oil companies have been strengthened and given new roles to play, and the desire to acquire a participating ownership share for the governments under the existing concession agreements is being vigorously pursued. In certain OPEC Member countries, various forms of production controls have been imposed.

REFINING

24.   The trend to construct refineries close to areas of consumption rather than close to sources of crude oil was further emphasized during the 1960s. Although the relative growth rate in refinery expansion was most striking in Japan, for every ton of capacity built in that country, four were added in Europe, as capacity grew from about 230 million tons in 1960 to 420 million tons in 1965 and to 700 million tons in 1970. Refinery development in North America has been comparatively modest, reflecting the very rapid enlargement programmes which had taken place prior to 1960.

25.   Refinery yield patterns vary considerably between OECD Europe, North America and Japan.   In Japan imports have been used to balance supply and demand; refiners have not been required to "balance the barrel". In North America, and especially in the United States, upgrading residual material to gasoline and middle distillates has been the main preoccupation, while the relative freedom to import fuel-oil has de-emphasized residual yields.   Finally, in Europe, refinery patterns have been influenced by location and by the composition of the readily available short-haul crudes from North Africa.

26.   It is thought that refinery capacity in Europe may almost double during the present decade and that in Japan the increase may be 2.5 times existing capacity.   Cost factors, concern for the environment and siting restrictions, coupled with pressures from the host governments, give considerable weight to the building of new refining capacity in the oil producing countries in the coming years.

TRANSPORT

27.   The most striking development in oil transport came in the latter part of the past decade, with the emergence of the VLCC (very large crude carrier).   The closure of the Suez Canal in 1967 and its continued closure since that time, the sharp rise in tanker building costs, and the reduced carrying costs implicit in the employment of the VLCC, all have combined to focus interest in the building of tankers 200,000 dwt and larger.

28.   Sea-borne movement of liquefied natural gas has also expanded dramatically.   It is thought that this specialized trade will grow rapidly in importance, along the lines of petroleum trade.

29.   The growth in seaborne movement of oil and the trend toward the VLCC have been, and will continue to be, important factors influencing the development of harbour facilities.   The limited number of deepwater ports capable of handling the VLCC has given rise to new patterns— transshipment, two-port discharge, lifting at sea and the like— of handling oil.

THE ENVIRONMENT

30.   Pollution abatement and the protection of the environment have become important social issues.   Pollution attributable to the operations of the petroleum industry is but a part of the total, but it is a highly visible part.   Pollution is an ever-present potential at each stage of the oil industry. But recognition of this potential and the implementation of safeguards at every step is a growing function of the oil industry.   In particular, allowable emissions from gasoline engines have been placed under rigid legislation, and a number of Conventions have been framed to minimize damage from oil pollution at sea.

CAPITAL REQUIREMENTS

31.   During the past decade the world oil industry, outside Eastern Europe, invested $157 billion in its effort to develop each stage of the

17

industry commensurate with the growth in demand. The need to expand oil production capacity attracted 43 per cent of the investment funds, although decreasing during the period as a whole; the construction of new refining and marketing facilities each accounted for 16 per cent; and investment in marine, pipe-lines, chemical plants and others accounted for the remainder. As might be expected, new investment in oil in the United States during 1960-70 came close to matching investment in the rest of the world (excluding Eastern Europe).

32.   The tremendous surge in oil demand anticipated for the 1970s—cumulative demand is likely to be double that of the 1960s— will require enormous amounts of capital. It is also likely that investment per unit of petroleum moving to market will be higher than in previous years. Overall financial needs could very well be in excess of $550 billion. This capital can come from the various provisions for capital recovery and from net income, and it can be borrowed. It is possible that the industry will have to generate an increasing share of the funds required from its own operations, which will require an attendant increase in net income.

NATURAL GAS

33.   The role of natural gas in the OECD was the subject of a detailed study published in 1969[1]. It can be noted here that natural gas is the most rapidly expanding sector among the individual forms of primary energy, and that its attractiveness as a clean non-polluting fuel and as a raw material for the petrochemical industry will further stimulate its growth in the years ahead.

OIL PRODUCT PRICES

34.   During the 1960s the prices (excluding taxes) for major oil products outside North America and Eastern Europe exhibited a declining trend. Technological progress and economies of scale in refining, declines in freight costs and plentiful crude oil supplies, all contributed to this decline. In 1970 and 1971, however, the rise in tanker rates and imposed crude oil cost increases for the producing companies led to oil product prices rising substantially throughout almost all of Europe and to a lesser extent in Japan. For a variety of reasons, including continued high growth rates in demand and enlarged competition for available crude oil supplies, it is unlikely that the stability in prices experienced during the 1960s will prevail during the 1970s.

---

1.   "Impact of Natural Gas on the Consumption of Energy", OECD Paris, 1969.

I

OIL, ENERGY AND THE OECD

INTRODUCTION

1.　The first chapter of this Report is devoted to mapping the growing use of oil in relation to other energy forms over the past decade and tentatively extending this view up to 1980.　Though special emphasis is given to the principal OECD areas— Europe, North America, Japan and Australia[1]—, broad indications are also furnished of developments in the rest of the world since any realistic survey of OECD energy requirements cannot be dissociated from global supply and demand trends.

2.　Forecasting energy requirements some ten years ahead is necessarily a somewhat arbitrary exercise requiring a series of basic assumptions— price trends, supply situation, economic growth and measures which could be taken for pollution abatement— which in the nature of things can never be accurately determined in advance and which each individually and jointly must influence significantly the extent and pattern of energy demand.

3.　Nevertheless, despite the imperfections and the likelihood that some or all the elements in the forecast will not develop exactly as foreseen, some view of the future is required, if only to outline possible paths of development and to introduce means of anticipation and contingency planning in the framing of energy policy.

4.　Arguably, four major assumption areas must be considered when developing a set of energy demand estimates of the form presented in this chapter:

　　i)　the expected rate of economic growth and its relationship to total energy demand;

　　ii)　price movements between energy forms and the global price of energy;

　　iii)　the supply situation;

　　iv)　measures which would be taken for pollution abatement.

5.　Though the four are necessarily interconnected, it is convenient to take each in turn to isolate the factors that require consideration.　First, the ratio of growth between energy requirements and Gross National

---

1.　See the Australian country chapter in Annex I for fuller coverage.

Product, which is considered to be a generally reliable guide forecasting[1], is itself the resultant of various opposing forces. Any trend towards increasing efficiency or economy in the use of energy (substitution of more efficent for less efficient fuels or appliances, the secular increase in the average efficiency of power stations, new process methods in industry lowering the input of energy per unit of output) will lower the ratio. Conversely, a trend towards more energy-intensive, as opposed to less energy-intensive, industries, substitution of energy for other factors of production (perhaps associated with a higher level of technology), growth in labour productivity generally, will have a contrary effect; as will the general rise in the standard of living leading to higher heating standards and the purchase of more energy-intensive appliances.

6.   A second factor is the trend in energy prices. The assumption of any shift in relative prices of substitutable energy forms will usually imply some substitution in the medium term. A rise in oil prices, for instance, not matched by equal increments for nuclear power would necessarily make the latter more attractive and hasten its development. Similarly, a rise or fall in the absolute price of energy in relation to prices of goods and services in general, would normally be expected to influence the use of energy as a whole. A relative hardening of energy prices would encourage development of less energy-intensive processes in industry, as well as the more rational use of energy in domestic and commercial sectors.

7.   Third, a forward view of demand must be related to supply possibilities. Indeed, supply limitations are a determining factor. For instance, it can be assumed that a market will exist for the full production potential of North Sea oil and gas within a wide cost range. Other aspects of supply can be determined for social reasons or for reasons of security of supply as is mostly the case with European coal production. Also demands for fuels can be and are affected by government policies on power stations, etc. These aspects must be considered in drawing up an integrated demand picture. For detailed assessment of current world oil reserves see chapter II.

8.   Finally, measures to be undertaken which hopefully will lead to pollution abatement and to preservation of the environment in themselves will require the expenditure of additional energy. A gallon of unleaded gasoline will not be able to do the work of a gallon of leaded gasoline; thus, all things being equal, an automobile will consume more fuel per kilometre travelled in the future when the lead content of gasoline is either reduced or eliminated entirely. The continued surge in demand for "clean" electric power for industrial and household use will have an accelerating effect on primary energy consumption, e.g. coal, fuel-oil and natural gas burned to generate this electric power. In the generation process, some two-thirds of the heating value of the primary fuel is lost, implying that a shift to conventional electric power is not the environmentally safe approach to our energy needs as may be thought. Cleaning up the lakes and rivers and the disposal of large accumulations of solid wastes are not only going to be expensive in monetary terms, but will involve a certain expense of energy.

---

1. The United States provided an exception to this general rule during the latter part of the 1960s (see paragraph 31).

9.   Generally the technique adopted for deriving the estimates in chapter I has been to pinpoint such elements as can be considered locally determined first (natural gas and coal in Europe, for instance) and to use macro-economic techniques (relationship of energy demand to GNP) for those parts that could not be thus determined (e.g. oil imports in Europe). Other basic assumptions have been pragmatic. It is assumed for instance that whilst the price of oil on the world market can be expected to rise in real terms during the next ten years, such price increases will not significantly curtail the vast market remaining for oil. Likewise, no radical changes for economic policy or other reasons are assumed that would dramatically upset existing energy patterns and relationships or the development of current trends. Again, for pragmatic reasons, single indicative totals are shown, rather than upper and lower ranges.

10.   A guiding principle in preparing the estimates has been to integrate expert view from a variety of bodies in OECD who are either directly concerned with energy problems or whose activities impinge upon them. Thus the basic economic assumptions adopted conform to the latest published work of the Economic Policy Committee[1], whilst the view of nuclear developments takes account of advice from the OECD Nuclear Energy Agency (NEA).

A. OIL IN OECD EUROPE'S ECONOMY

THE ENERGY PATTERN 1960-1970

Total requirements and the economy

11.   Total primary energy requirements in OECD Europe grew from 612 million tons of oil equivalent (Mtoe) in 1960 to 1,041 Mtoe in 1970, i.e. at an average annual rate of some 5.5 per cent with, of course, considerable differences between individual Member countries (annual growth has ranged from less than 2 per cent in the United Kingdom to more than 10 per cent in Finland, Greece and Italy). This growth has not followed a regular pattern. Rates above 7 per cent prevailed in the early 1960s, but dropped substantially in the middle 1960s and have now risen again to high levels in the last three years.

12.   The relationships between trends in energy requirements and economic growth (as measured by GNP) have remained generally above unity throughout the period 1960 to 1970. The growth in GNP exceeded that of energy requirements for only three years of the decade— 1961, 1964 and 1966. The average rate of growth in GNP was some 4.8 per cent per annum.

The share of the various forms and sources of energy in total demand

13.   The past decade has been marked by a decline in the use of coal, the rapid growth in that of oil, and to a lesser extent, the development of natural gas production and of nuclear power.

---

1. "The Growth of Output 1960-1980", OECD, December 1970.

14. Solid fuels provided some 60 per cent of total European energy requirements in 1960. By 1970 this contribution had fallen to below 30 per cent. The decline has been both relative and absolute, as solid fuel consumption fell from 375 Mtoe in 1960 to 306 Mtoe in 1970 (i.e. from 535 to 435 million tons in terms of coal equivalent).

15. In an expanding energy market, coal and, to a lesser extent, lignite, have not been able to maintain their general position and for reasons of price and of convenience have consistently lost ground. An absolute decline in demand of some 90 Mtoe among the final consumption sectors[1] has only partially been compensated over the period by gains in the electricity market.

16. The principal competitor has been oil. Oil has increased its share in total energy demand from 33 per cent in 1960 (199 Mtoe) to 60 per cent in 1970 (620 Mtoe). By far the greater part of this total increase in demand has had to be met by imports, since indigenous oil production has only risen marginally over the period from 16 million tons in 1960 to 23 million tons in 1970 (8 per cent and 4 per cent of total oil demand respectively).

17. Substantial new developments have occurred in natural gas production and imports and in nuclear power generation, but the contributions from these sources, though growing fast, are still minor in overall terms. Natural gas and nuclear power together provided some 7 per cent of total energy needs in 1970.

18. Hydro power has increased its absolute contribution to European energy requirements, but in percentage terms has gradually been losing ground due to the limited scope for harnessing new hydro resources in Europe. In 1970 it provided 3 per cent of total energy requirements.

19. Geothermal energy is in practice currently limited to Italy, and over the period has provided some 0.3 Mtoe annually.

The position of oil and the product market

20. During the years 1960-70, demand for oil, increasing at an average rate of 12 per cent per annum, moved into first place as Europe's major energy supplier, overtaking solid fuels, the traditional European energy source, by the middle of the decade. Its price advantage, its convenience in transport and storage and in use, and above all its plentiful supply, all contributed to the present dominance of oil on the European energy market. Oil in its various product ranges has penetrated all sectors of the market, but especially the final consumption sectors.

21. Oil increased its share of the domestic and miscellaneous market from 23 per cent in 1960 to 55 per cent in 1970. Its penetration in industry (45 per cent) is of somewhat lower magnitude. For transport,

---

1. In this and subsequent paragraphs, final consumption sectors mean the final users of primary and secondary energy forms. Final users, for example, are industry, agriculture and households. Primary and secondary energy forms include, among others, coal, coke, patent fuel, liquid fuels and electricity.

coal as a fuel has by now been all but completely ousted, and oil in practical terms is the only significant fuel. The share of oil as a power station fuel rose from 9 per cent in 1960 to more than 25 per cent in 1970. The inroads of oil in this sector, where solid fuels still retain their hold, have been substantial but nevertheless below performance elsewhere. This is partly because in certain large energy-consuming countries having indigenous coal industries, oil penetration has deliberately been checked for social and energy policy reasons.

22.   Table 3 shows the change in the pattern of consumption of the principal oil products in OECD Europe over the decade 1960 to 1970. The general pattern has been towards a relative increase in the share of the more valuable lighter products, with the exception of gasoline, whilst the share of residual products remained more or less stable. With a 40 per cent share of the total market, residual fuel-oil has a clear dominance over any other group of products. The aviation fuel, kerosenes and other product groups show the highest rate of growth, just slightly above that for gas/diesel-oil, but this is in large part due to the inclusion of naphtha for which a substantial market has developed. The table shows that demand for naphtha rose from 17 million tons in 1967 to 27 million tons in 1970.

23.   The current growth in demand for naphtha and other oil products for non-energy purposes, mainly petro-chemicals (see chapter V, Refining) is a significant new development in the product market. From a total of 12 million tons in 1960, this market took some 54 million tons in 1970. This represents an annual growth rate of about 16 per cent, i.e. one and a third times the overall growth rate for oil. The market for non-energy products for transport and for bunkers together is that part of oil demand which can be considered as specific, i.e. where in the short and medium term effective substitution by other forms of energy is precluded. Over the period there was a relative decrease in European reliance on oil for specific needs. Oil requirements in this category fell from 43 per cent of total oil demand in 1960 to about 35 per cent in 1970.

OECD EUROPE: OIL DEMAND IN 1980

24.   Based on country-by-country estimates and applying rates of economic growth as published in the OECD report "The Growth of Output 1960 to 1980"[1], it would seem that overall energy requirements in 1980 could reach 1,760 Mtoe, of which 63 per cent could be oil. The implied growth rate (5.4 per cent per annum over the period 1970 to 1980) is above that assumed for economic growth (4.8 per cent per annum) and reflects the view that at least over the next decade the trend to energy intensiveness in the economy as a whole will outweigh any increase in efficiency of use of energy or movements to contain energy demand.

25.   In terms of energy consumption per capita, the estimates imply an increase from some 3 t of oil equivalent in 1970 to about 4.5 toe per capita in 1980. This can be compared with a likely level of 5.5 toe in Japan and 11 toe in North America by the end of the decade.

---

1. OECD, 1970.

26.   In meeting total requirements, the contribution from indigenous oil and gas resources is assumed to be higher than ever before. This can be ascribed to the rising output from the Dutch gas field at Slochteren as well as the newly-found reserves within Europe's Continental Shelf. By 1980, the United Kingdom and Norway can be expected to become major oil and natural gas producing countries. Prospects for discoveries offshore of Spain are also promising. In quantitative terms, estimates suggest that indigenous oil and natural gas will be meeting more than 20 per cent of total energy requirements by 1980. This is over twice the 1970 level. Underlying this estimate is the view that North Sea oil output may well be approaching 150 million tons by the end of the decade.

27.   These developments are likely to cut the rate of increase in oil imports by almost two-thirds. Over the period 1960 to 1970, oil imports rose by 12.5 per cent per annum, but this growth rate could drop to some 4.5 per cent between 1970 and 1980. This drop can be expected to have a beneficial effect on the balance of payments of the principal European producing countries as well as improving security of supply for Europe as a whole. Nevertheless the import total itself would in absolute terms remain high and in 1980 could approach 950 million tons (see chapter II, B, Obtaining Reserves in the Future).

28.   In 1980, hydrocarbons can be expected to retain their present clear dominance of the European energy market. Oil and natural gas together would be likely to cover over three-quarters of total energy demand. Government policies could have a significant effect on demands for various fuels, particularly coal, and though coal is expected to continue losing ground under pressure from oil and nuclear power, a wide range of uncertainty affects this and other figures. By 1980 coal could be providing less than 14 per cent of total energy needs as opposed to 29 per cent in 1970. However, it should be noted that this global decline masks certain contrary tendencies in the coal market. Loss of markets is expected to be limited to the substitutable segment of coal demand (i.e. in final consumption sectors and to a lesser extent in power stations) whilst the demand for metallurgical coking coal can be expected to rise. The coking coal market represented some 25 per cent of total solid fuel consumption in 1960, rising to 30 per cent in 1970. A recent report published in OECD[1] foresees regular increases in this market matching an expanding output from the iron and steel industry. Coal for coking purposes can thus be expected to further increase its overall share in total coal demand, perhaps approaching a 40 per cent share of the solid fuel market by 1980.

29.   Given the limited scope still available in Europe for harnessing hydro resources, this energy source is expected to show only moderate development over the current decade. Substantial growth, however, is predicted for nuclear power which by 1980 could be meeting 7 per cent of total energy requirements. This estimate is based on the realization of current programme plans. It assumes that no substantial contribution from the breeder reactor can be expected over the next decade.

---

1.   "Report on the Problems and Prospects in the Coking Industry in the OECD Countries", OECD, 1972.

30.   Over the period to 1980, the pressures to contain atmospheric pollution and thermal pollution resulting from increased levels of fuel combustion are likely to gain momentum.   This will be reflected in a marked demand which may raise questions of scarcity for low sulphur oils and smokeless fuels and will raise new problems where energy and environmental policies will have to be reconciled[1].

B.   OIL IN THE NORTH AMERICAN ECONOMY

THE ENERGY POSITION 1960-1970

Total requirements and the economy

31.   Total primary energy requirements in North America grew from 1,138 Mtoe in 1960, to 1,762 Mtoe in 1970, i.e. at an average annual rate of 4.5 per cent.   The rate of increase in GNP over the same period was some 4.1 per cent per annum, on average.   Whereas in Japan and Europe the relationship between energy requirements and GNP was generally a consistent one with a predictable correspondence between movements in economic growth and the demand for energy, this has not been the case in North America.

32.   Over the first half of the decade for North America, the ratio of primary energy to GNP was a falling one.   At mid-decade this curve flattened out and in the latter half of the decade began to rise steeply.   This upturn has been attributed to the increased use of air conditioning, greater use of electrical processes in industry, and a decrease in opportunities to improve efficiencies in energy use.

The share of the various forms and sources of energy in total demand

33.   In contrast to Europe and Japan, the share of oil in total demand has decreased marginally over the period from 45 per cent in 1960 to 43 per cent in 1970.   It was, however, the single major fuel source for the decade and the growth in absolute terms showed an overall increase of some 4 per cent per annum.   (Included in oil are natural gas liquids which represented 63 million tons of oil equivalent in 1970.   Over the decade they increased their share in total oil demand from 7 per cent in 1960 to 8.3 per cent in 1970.)

34.   Coal, holding about half the share of oil on the energy market, also declined in relative importance.   From 23 per cent in 1960, coal retained 20 per cent in 1970.   Yet, in volumetric terms, demand for coal increased at just under 3 per cent per annum.   The specific requirements for coking coal for production of coke grew at about half this rate and thus declined as a share of total coal demand.   From 21 per cent in 1960, their share fell to 19 per cent in 1970 and is expected to further decline during the 1970s.   Mention should also be made of the export market for coking coal.   Exports of coking coal of 37 million tons from the United States

---

1. See chapter VII.

and 1 million tons from Canada in 1969 are likely to reach 40-44 and 12-16 million tons respectively by 1975[1].

35. The energy form exhibiting the highest rate of growth over the period was natural gas. Demand for gas grew at some 6 per cent per annum with its overall market share rising from 29 per cent in 1960 to 34 per cent in 1970.

36. Hydro power and small amounts of geothermal energy maintained a steady contribution of around 2.5 per cent over the period. Nuclear energy which was insignificant in 1960 contributed 6 Mtoe in 1970. This is just over half in absolute terms of the contribution of nuclear power to European energy requirements in the same year.

The position of oil and the product market

37. The position of oil on the North American market is influenced by the fact that over three-quarters of total consumption is covered by indigenous production. Unlike Europe and Japan, low-price imports are controlled for energy policy reasons. North America also has its own substantial reserves of coal and natural gas which compete directly with indigenous oil. The net effect is to reduce overall dependance on oil which in 1970 provided some 43 per cent of total requirements, as against 60 per cent in Europe and over 70 per cent in Japan.

38. The position of Canada, whose overall share of North American energy demand in 1970 was just under 8 per cent, differs markedly from that of North America as a whole. Coal and natural gas have a much smaller share of the market (14 per cent and 23 per cent respectively in 1970) whilst the share of oil in total demand (51 per cent in 1970) is higher. This should be seen against the background of Canada as a net energy exporting country. In 1970 its imports of oil and coal were offset by substantial exports of natural gas and oil to the United States and a growing coal trade with Japan.

39. During the 1960s the share of oil in total sector demand in North America fell from 47 per cent in 1960 to 45 per cent in 1970. This level of penetration for oil is very much less than in Europe and Japan and reflects the substantial contributions from natural gas. In sector demands as a whole, natural gas increased its share from 33 per cent in 1960 to 38 per cent in 1970. Only in the transport sector where oil has no clear competitor (60 per cent of all final consumption of oil is in this sector) was it able to maintain its position over the period. Transport as a whole represented some 27 per cent of total sector demands between 1960 and 1970.

40. Oil contributed to electric power generation requirements in a relatively minor way. In 1960 oil provided 7 per cent of total inputs in North America, rising to 12 per cent in 1970. This market has been very much the preserve of natural gas and coal; these fuels in 1970 provided 25 per cent and 52 per cent respectively of total inputs.

1. "Report on the Problems and Prospects in the Coking Industry in the OECD countries", OECD, 1972.

41. Of total oil requirements of 763 Mtoe in 1970, 82 per cent went into final consumption, 7 per cent into power stations, 10 per cent into non-energy products, with the remainder being losses. Of this total, the specific element (transport, bunkers and non-energy products) slightly increased its share of total oil demand from 57 per cent in 1960 to 58 per cent in 1970. The fastest growing element in this total, especially in the earlier part of the decade, was the demand for non-energy products which rose at an average of 6 per cent per annum, i.e. over the period, demand for oil for non-energy purposes showed one and a half times the growth rate of oil in general.

42. The demand for petroleum liquids as raw material for the manufacture of petrochemical derivatives and other so-called non-energy (or non-fuel) uses is a small but rapidly growing share of the total demand for petroleum in the United States. Petrochemical feedstocks account for approximately one-half of these non-energy uses; asphalt and road oil demand, reflecting the continuing interstate highway construction, make up some 25 per cent of the total; lubricants account for something less than 10 per cent. The remainder takes into consideration such diverse applications as coke used in the making of electrodes, and waxes.

43. Table 5 shows the breakdown of oil demand by major product category for North America over the period 1961 to 1970. The pattern is very different from that of Japan and Europe, with a very large preponderance given to white products. In 1970, fuel-oil represented only 18 per cent of total North American demand for products as opposed to 40 per cent in Europe and 50 per cent in Japan. The single largest product group is gasoline, which represented 35 per cent of all products in 1970.

44. This situation has prevailed throughout the period, with some fluctuations in the relationship between products. Though still largely preponderant, the share of gasoline has in fact fallen over the period, whilst fuel-oil has moved from a 16 per cent share of the market in 1961 to 18 per cent in 1970. The share of gas/diesel-oil has declined from 22 per cent in 1961 to 19 per cent in 1970, whilst the aviation fuel, kerosene, and naphtha group has increased its share. The naphtha market has remained virtually stable at some 20 million tons per annum for the period 1967 to 1970. As can be seen from table 5, special naphtha use for solvents recorded the largest growth within the naphtha market, but this growth has been offset by the shift from naphtha jet fuel consumption to a kerosene-type jet fuel.

OIL DEMAND IN 1980

45. Total energy demand in 1980 is estimated at 2,870 Mtoe, i.e. increasing at an average annual rate of 5 per cent which is above the level experienced in the 1960s. A continued expansion of the North American economy, greater use of electricity, and an increased rate of energy consumption per unit of GNP account for the projected higher energy growth rate in the 1970-80 period.

46. Energy consumption per capita is considerably higher in North America than in the other OECD areas. Indeed, consumption per capita

27

in 1960 (5.7 toe) was already higher than either of the estimated levels for Japan and Europe in 1980. From a level of 7.8 toe per head in 1970, consumption is expected to reach 11.2 toe in 1980.

47.   Oil demand in North America could reach 1,200 million tons by 1980, of which a larger part than heretofore would be provided by imports. Imports as a share of total oil demand are expected to increase from 22 per cent in 1970 to nearly 40 per cent in 1980 (see chapter III, Sources of Oil Supplies).

48.   The growth in demand for oil for the period 1970-80 is expected to be higher than in the previous decade (4.7 per cent as opposed to 4.1 per cent per annum). This rate of growth would imply that oil will largely maintain its present share of the energy market, i.e. it would continue to provide more than 40 per cent of total energy demand. Coal is expected to experience a small decline in its market share (18 per cent as opposed to 20 per cent in 1970) as would natural gas (34 per cent down to 30 per cent). The sector to increase its market share is that grouping hydro and nuclear power. The latter, notably, is expected to reach substantial proportions in the 1970s, perhaps contributing the equivalent of 230 million tons of oil in overall requirements in 1980 (i.e. about twice the level forecast for Europe).

49.   As in the other OECD areas, environmental considerations are expected to gain in importance over the coming decade. Recent work in OECD suggests that one difference between the North American and European or Japanese situation is that assuming no additional abatement measures over and above those already planned, coal, rather than oil combustion, would in 1980 be responsible for the bulk of sulphur emissions. This reflects the high sulphur content of suitably located North American coal reserves. Environmental considerations are examined further in chapter VII.

C.   OIL IN JAPAN'S ECONOMY

THE ENERGY CONSUMPTION PATTERN 1960-1970

Total requirements and the economy

50.   Over the decade total primary demand for energy increased in Japan from 86 Mtoe in 1960 to 266 Mtoe in 1970, which represents an annual rate of increase of just under 12 per cent. This very high rate of growth in energy requirements has more than kept pace with the overall growth in the economy which, on average, has shown an 11 per cent annual increase over the period. Growth in energy demand has not been at a consistently high level. In practice, one year of lower growth has been followed by two years with very high rates. The low years in the decade were 1962, 1965 and 1968.

The share of the various forms and sources of energy in total demand

51.   As in Europe, Japan's energy structure has been marked by a relative decline in coal and a high growth in oil demand; it is the latter

28

that has provided the bulk of the overall increase in energy requirements. The contributions of other forms of energy— hydro power, natural gas, nuclear power (from 1966)— are slight.

52.   Coal and small quantities of lignite represented together 55 per cent of total demand in 1960.   By 1970, the share of these fuels had fallen to 23 per cent.   Nevertheless, in contrast to Europe, this did not represent an absolute decline, inasmuch as consumption of coal and lignite rose by 30 per cent over the period, from 47 Mtoe in 1960 to 62 Mtoe in 1970, due mainly to the thriving activity of the iron and steel industry.   Coke oven coal demand increased from 13 million tons in 1960 (19 per cent of total coal requirements) to 52 million tons in 1970 (59 per cent of total requirements).

53.   Oil, of which all but 1 million tons is imported, increased its share of the market from 36 per cent in 1960 to 72 per cent in 1970.   The compound growth rate has been 20 per cent per annum.   Imported oil plus imported coal has raised the overall dependence of the economy on fuel imports from 43 per cent in 1960 to 85 per cent in 1970.

54.   Small quantities of natural gas are produced in Japan and with the first imports represented some 1 per cent of total demand in 1970. Hydro power has consistently provided some 8 Mtoe to overall requirements during the period.   A newcomer on the energy market is nuclear power which entered production in 1966.   The contribution from this source is building up rapidly, but has not substantially influenced the Japanese energy market of the past decade.

The position of oil and the product market

55.   The high rate of increase in oil demand has been largely absorbed in the final consumption market.   From 35 per cent in 1960, oil increased its share of this market to 63 per cent in 1970.   The principal gains were in the domestic and miscellaneous sector.   The increase in oil demand over the period 1960 to 1970 of some 160 million tons has been distributed along these lines: 70 per cent to final consumption, 20 per cent to power stations, 10 per cent to non-energy purposes and refinery and other losses.

56.   Table 7 shows the pattern of demand for oil products in Japan for the period 1960 to 1970.   The relative share of fuel-oil has been declining, falling from nearly 60 per cent to 51 per cent of total products over the period.   The relative shares of gasoline and gas/diesel-oil have also declined.   The group to increase its share of the market is that which includes aviation fuels, kerosenes, naphtha and other products.

57.   One rapidly expanding market is that for non-energy oil products (see chapter V, Refining) whose consumption increased from 1.8 Mtoe in 1960 to 17.9 Mtoe in 1970.   However, added to oil demand for bunkers and transport, this category saw its share of total oil demand decrease from 34 per cent in 1960 to 28 per cent in 1970.   This trend to reduced reliance on oil in specific applications matches a similar trand experienced in Europe and is in contrast to the situation in North America where this reliance increased slightly.

58.   The latest estimates of Japanese energy requirements in 1980 foresee a total demand of around 635 Mtoe.   This implies a growth rate in energy demand of just over 9 per cent per annum over the period 1970 to 1980, i.e. below the level for economic growth which is expected to grow at 9.5 per cent per annum.

59.   This reduction in the ratio between energy demand and GNP compared with the previous decade may be seen as a reaction to the very high rates of growth already experienced and being experienced which, for reasons of rational use of energy and protection of the environment, could suffer a downwards pressure in future years.   The high density of population, the reliance on high sulphur crudes and the rapid progress in the economy mean that environmental problems are particularly severe in Japan and steps to limit pollution are to be expected over the present decade.

60.   Japan has experienced a particularly rapid increase in energy consumption per capita.   From 0.9 toe in 1960 (i.e. at about the average world level) per capita energy consumption increased to some 2.5 toe in 1970 and could reach 5.5 toe by 1980, thus overtaking the likely consumption level for Europe.

61.   The dependence on oil and with it a dependence on imports are expected to be higher in Japan at the end of the decade than in the other OECD areas.   Despite a halving in the expected rate of increase in oil demand, the share of oil in total demand could reach 73 per cent by 1980 (466 Mtoe).   The greater part would have to be imported, inasmuch as indigenous oil production is not expected to much exceed 2 million tons, or 0.4 per cent of total energy demand.   Other imports— coal, mostly coking coal, and substantial quantities of liquefied natural gas— may raise the total dependence on imported energy to nearly 90 per cent of total requirements.

62.   Hydro power is expected to show only a marginal increase, but substantial growth in the contribution from nuclear power is anticipated. This latter source of energy could be contributing some 6 per cent of total requirements by 1980 (the equivalent of some 40 million tons of oil).

D.   OIL IN AUSTRALIA'S ECONOMY

THE SHARE OF THE VARIOUS FORMS AND SOURCES OF ENERGY IN TOTAL DEMAND

63.   The past decade in Australia has been marked by considerable change in the energy balance.   The main sources of energy were oil and coal, both black coal and lignite.   Consumption of coal increased in terms of actual tonnage but the proportion of the total energy derived from coal decreased due to the strong growth in oil demand.   Significant and widespread natural gas discoveries were made towards the end of the decade and capital cities in three States started to use this fuel in 1969.

64.   Oil supplied 37 per cent of the primary energy market in Australia in 1960 and had increased its share to almost 50 per cent in 1970. It is expected to maintain this proportion at least until 1980. Growth in oil consumption averaged 7 per cent per annum from 1955-1960, 9 per cent from 1960-65 and 7.7 per cent from 1965-1970. Annual average growth for the next decade is expected to be 6.4 per cent.

65.   Fuels for transport comprised about 60 per cent of oil demand in 1960. The figure had fallen to around 53 per cent in 1970. In 1970 approximately 28 per cent of primary energy consumed went into electricity generation. Approximately 360,000 t of oil were used for electricity generation in 1960. In 1970 usage was 496,000 t or only 3.5 per cent of primary energy consumed in electricity generation.

66.   The overall share of coal (black and brown) in the energy market weakened over the decade from 55 per cent to 43 per cent. The share of black coal dropped from 42 per cent to 31 per cent. Over the period some traditional markets for black coal declined as the railways turned to diesels and reformed petroleum fractions began to replace coal in gas making.

67.   The average annual increase in black coal consumption from 1960-70 was only 2.2 per cent. The increase through to 1980 is expected to average 6.2 per cent which will keep black coal's share of the total market around 30 per cent. The growth will take place in the electricity and the iron and steel industries. In 1970, 80 per cent of black coal was used in these two industries.

68.   Electricity generation in Victoria is based on vast brown coal deposits in the Latrobe Valley. In 1970, 76 per cent of brown coal mined was used for public authority electricity generation. Although its share of the energy market is falling slowly brown coal will still have about 9 per cent of the energy market in 1980.

69.   Australia is not well endowed with potential hydro power. The main developments are in Tasmania and the Snowy Mountains area. Output of hydro-electricity is increasing but this source of energy accounted for only 2 per cent of the primary energy market in 1970 and its share will fall below this by 1980.

70.   Wood is still used as a fuel in a small way in some factories (mainly in country areas) and for domestic use. However, it is rapidly being replaced by more convenient fuels. The Mount Gambier power station is fuelled by wood waste (186,000 t in 1970) from an adjacent sawmill and this is likely to continue. Bagasse, a fibrous residue from the sugar cane industry in Queensland and New South Wales, is used to raise steam and generate electricity. These two fuels provided about 3.5 per cent of primary energy used in 1970 compared with 6.7 per cent in 1960.

71.   The position of oil up to 1980 is discussed in the country chapter in Annex 1, under the headings "Growth of Inland Consumption of Main Oil Products" and "Oil Consumption in relation to Total Energy Consumption".

E. OIL AND THE REST OF THE WORLD[1]

72.   Given that OECD energy requirements cannot be dissociated from the total supply and demand situation and need to be set against some view of world trends, an attempt is made in the following paragraphs to develop broad indications of possible developments in energy demand[2] in the rest of the world.   These estimates are based on a variety of sources both internal and external to OECD[3].   Whilst it is believed that these estimates give a fair view of likely developments, they are not to be regarded as a formal assessment and should be considered as indicative only.   Accordingly, no attempt is made in the text to provide detailed arguments for the figures given; rather, the comments themselves are limited to a description of summary trends.

73.   Taking the rest of the world as a whole, oil held 22 per cent of total requirements in 1960, increasing to 32 per cent in 1970.   Projections for 1980 foresee a further increase in the share of oil, which could reach 39 per cent of total requirements.   In quantity terms, oil demand has grown from 274 million tons of oil in 1960 to 590 million tons in 1970, rising to a possible total of 1,260 million tons in 1980.   Over the period 1960 to 1970, natural gas was the fastest growing fuel (14 per cent of total requirements in 1970) with coal the slowest.   Indeed the share of coal in total demand fell from 70 per cent in 1960 to 53 per cent in 1970.   The share of coal is projected to continue falling relative to other forms of energy and could be providing just over 40 per cent of total demand in 1980 (1,300 Mtoe).   The contribution from hydro power and nuclear power is not expected to reach substantial levels and could be providing only 2 per cent of total requirements in 1980.   Overall requirements could rise at a faster rate over the decade 1970 to 1980 than in the previous period (5.7 per cent as opposed to 4.2 per cent) reaching a level of over 3,200 Mtoe in 1980.   A growth rate of 5.7 per cent would probably be higher than that experienced within the OECD area with a consequent (though slight) decrease in the OECD area's share of world energy demand.

74.   Among the five main regions which make up the rest of the world (Oceania, West Asia and the Far East, Africa, South America, Eastern Europe and China), the region showing the fastest growth in energy requirements both for the period 1960 to 1970 and in the projections to 1980 is West Asia and the Far East.   The region with the highest level of energy

---

1.   Oceania, West Asia and the Far East, Africa, South America, Eastern Europe and China: United States Delegation to OECD; Remaining regions: based
2.   The historical data and estimates relate only to commercial forms of energy. In practice large quantities of wood, dung and the like are consumed as fuel in these countries.   Substitution of commercial energy forms for non-commercial forms will distort the apparent growth trend in energy demand.   The growth rate will appear to rise faster than is in fact the case in real terms.
3.   Sources.   Oceania: largely based on Australian country chapter; Eastern Europe and China: United States Delegation to OECD, Remaining regions: based upon projections contained in "Middle Eastern Oil and the Western World - Prospects and Problems", Schurr and Homan, New York, 1971.

requirements is Eastern Europe and China, which in 1970 accounted for just under three-quarters of the total demand. The general pattern for each of the regions is one of higher rates of growth in energy requirements in the 1970s.

SUMMARY

OIL AND WORLD ENERGY DEMAND

WORLD CONSUMPTION TRENDS: 1960-1970

75.   The combined energy requirements for the three main OECD areas— Europe, North America and Japan— plus the rest of the world, equalled some 3,060 million tons of oil equivalent (Mtoe) in 1960, rising to 4,920 Mtoe in 1970. The demand for energy among these countries expanded at an annual average rate of 4.9 per cent during the past decade, and represented an average consumption per capita in 1970 of nearly 1.5 t of oil equivalent.

76.   Among the various forms of primary fuels, oil increased its share from 33 per cent in 1960 to 44 per cent in 1970. Whereas at the start of the decade, coal was the major world fuel, oil overtook coal towards the middle of the decade and now holds this premier position.

77.   Far behind oil in overall world significance, but yet showing the highest growth rate over the period is natural gas, which raised its contribution to overall requirements from 14 per cent in 1960 to 19 per cent in 1970.

78.   Hydro power (and the limited quantities of nuclear power) grew at an annual average rate of 6 per cent per annum and claimed in 1970 some 3 per cent of the world energy market.

WORLD CONSUMPTION TRENDS: 1970-1980

79.   World energy consumption is projected to increase over the present decade from 4,920 Mtoe in 1970 to some 8,480 Mtoe in 1980, i.e. at an average annual rate of 5.6 per cent, which would exceed the rate experienced between 1960 and 1970 (world consumption per capita in 1980 could reach 1.9 toe). This increased consumption rate reflects the assumption of an accelerating demand for energy in the "rest of the world".

80.   Of the total energy demand projected for 1980, oil could claim some 48 per cent (exceeding 4 billion tons), coal some 25 per cent, gas 20 per cent, with the remainder made up of hydro and nuclear power. The latter is expected to be the fastest growing element (16 per cent per annum), largely due to growth in the OECD areas, especially North America.

81.   A significant trend for the 1970s is the higher growth in demand for solid fuels. From an almost stationary 1 per cent per annum growth experienced in the 1960s, projections foresee an annual increment of just over 2 per cent in the 1970s. Only in Europe is the demand for coal expected to show further decline during the 1970s.

33

82.   Flanking the higher growth in demand for solid fuels and in nuclear power, lower growth rates in the present decade are contemplated for oil and natural gas. Nonetheless these two fuels together will still largely dominate the world energy market in 1980 and may provide two-thirds of total requirements.

OECD AREAS IN RELATION TO WORLD TRENDS

83.   The share of the OECD areas as a whole in world energy demand, which increased slightly during the period 1960 to 1970, from 60 per cent to just over 62 per cent, could show a slight decline by 1980 as the level of demand in the rest of the world moves ahead. However, in terms of consumption per capita the OECD areas are expected to continue showing a striking superiority compared to the rest of the world. Whereas in the rest of the world consumption could rise from 0.6 t of oil equivalent per capita in 1970 (0.5 in 1960) to just under 1 toe per capita in 1980, the OECD area as a whole could reach a level of nearly 7 toe per capita in 1980 as compared with 4.5 toe in 1970 and some 3 toe in 1960. This situation is summarized in the following table.

ENERGY CONSUMPTION PER CAPITA PER ANNUM[1]

	1960	1970	1980
North America	5.7	7.8	11.2
Japan	0.9	2.6	5.5
Europe	1.8	2.8	4.4
Total OECD	2.9	4.4	6.8
Rest of World	0.5	0.6	0.9
Total World	1.0	1.35	1.9

In tons of oil equivalent.

1. Source of population data: "Enquiry into Demographic Trends in Member countries", OECD Working Document, 1971; UN Demographic Yearbook; UN Monthly Bulletin of Statistics, November 1971.

CONCLUSIONS

84.   Our forecasts underscore a continuing demand world-wide for ever-growing quantities of oil. It is expected that by 1980 world oil demand (at some 4 billion tons) would be twice the demand on an equivalent heat basis for any other form of energy. Of this demand for oil, some 70 per cent may originate in the OECD area. The quantities involved and the expectation of further growth dictate the importance of a full review of the role oil plays in the OECD Member countries. Such a review, which covers anticipated developments and requirements as well, is attempted in the chapters which follow.

Figure 1

PRIMARY ENERGY REQUIREMENTS IN OECD EUROPE
(1960 to 1980)

Million tons of oil equivalent

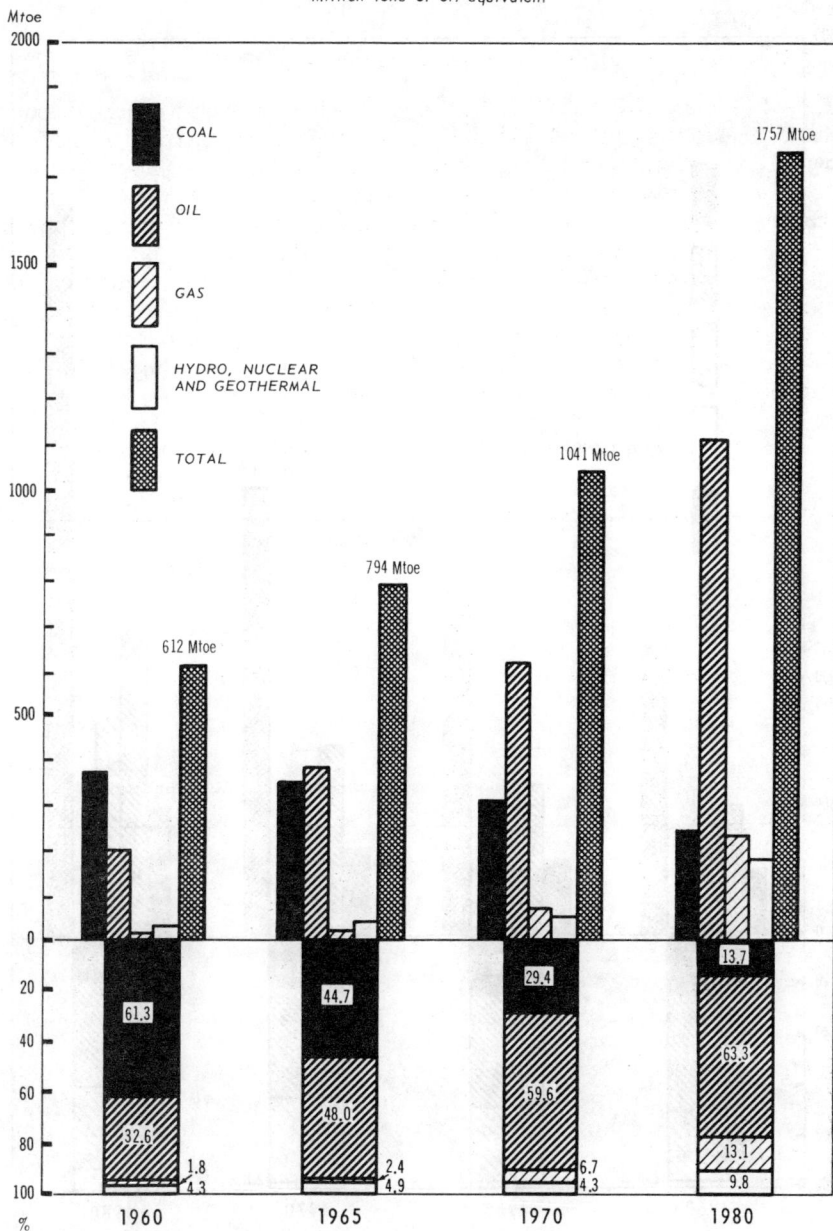

Figure 2

PRIMARY ENERGY REQUIREMENTS IN NORTH AMERICA

(1960 to 1980)

Million tons of oil equivalent

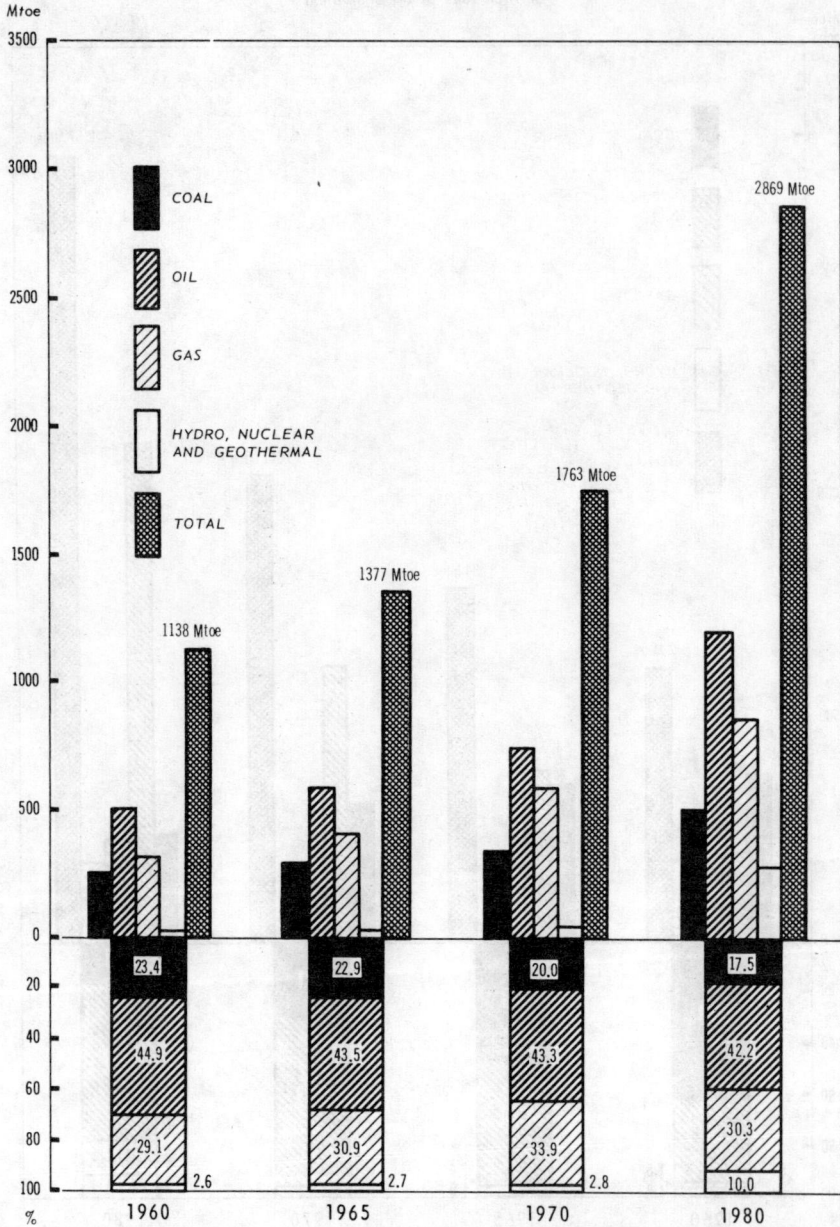

Figure 3
PRIMARY ENERGY REQUIREMENTS IN JAPAN
(1960 to 1980)
Million tons of oil equivalent

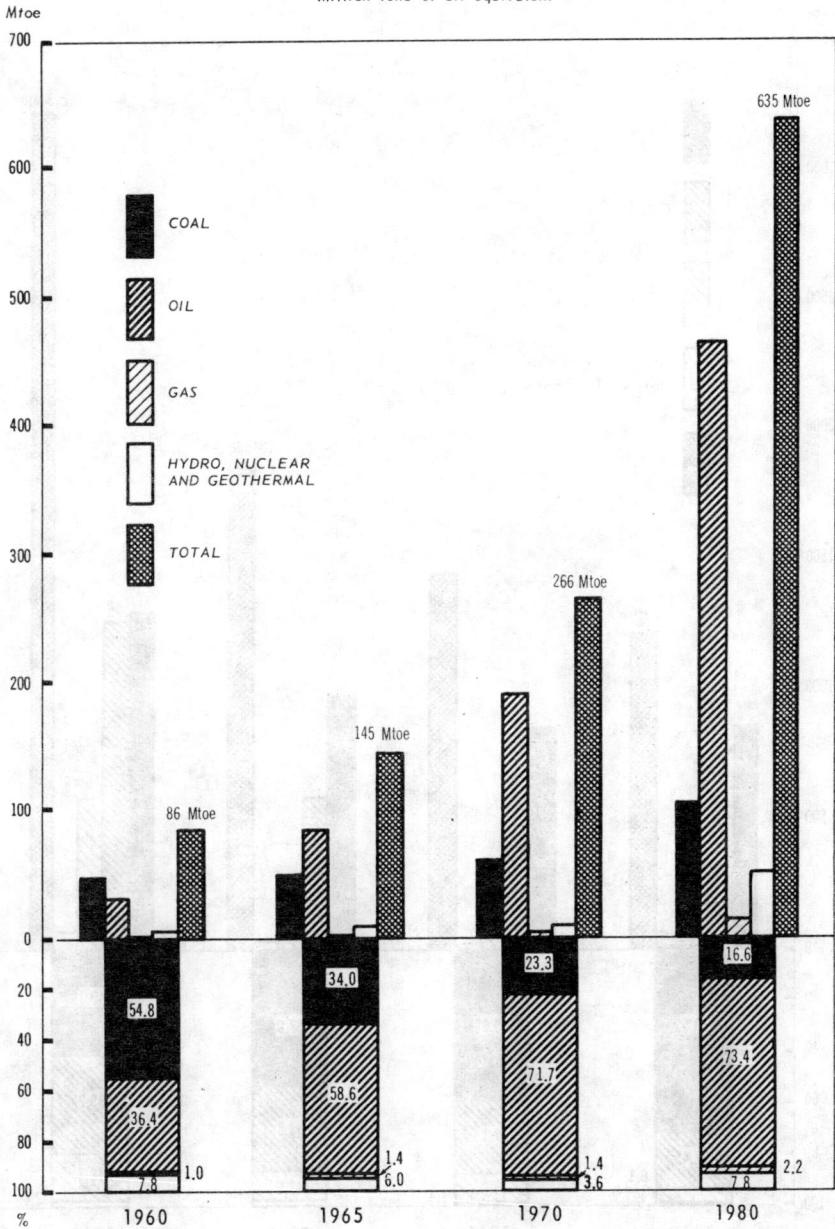

Legend:
- COAL
- OIL
- GAS
- HYDRO, NUCLEAR AND GEOTHERMAL
- TOTAL

Mtoe

Year	COAL	OIL	GAS	HYDRO	TOTAL
1960	54.8	36.4	1.0	7.8	86 Mtoe
1965	34.0	58.6	6.0	1.4	145 Mtoe
1970	23.3	71.7	3.6	1.4	266 Mtoe
1980	16.6	73.4	2.2	7.8	635 Mtoe

%

37

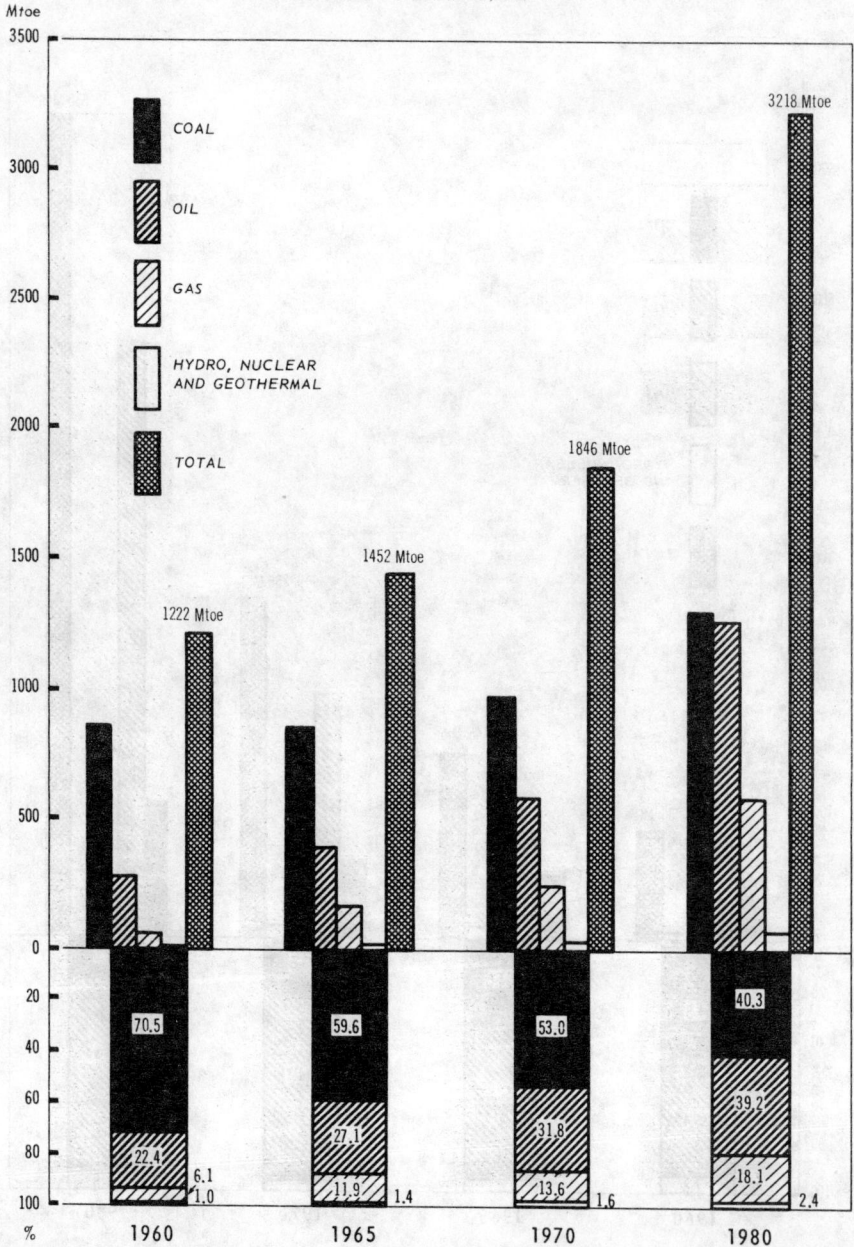

Figure 4

PRIMARY ENERGY REQUIREMENTS IN THE REST OF THE WORLD
(1960 to 1980)

Million tons of oil equivalent

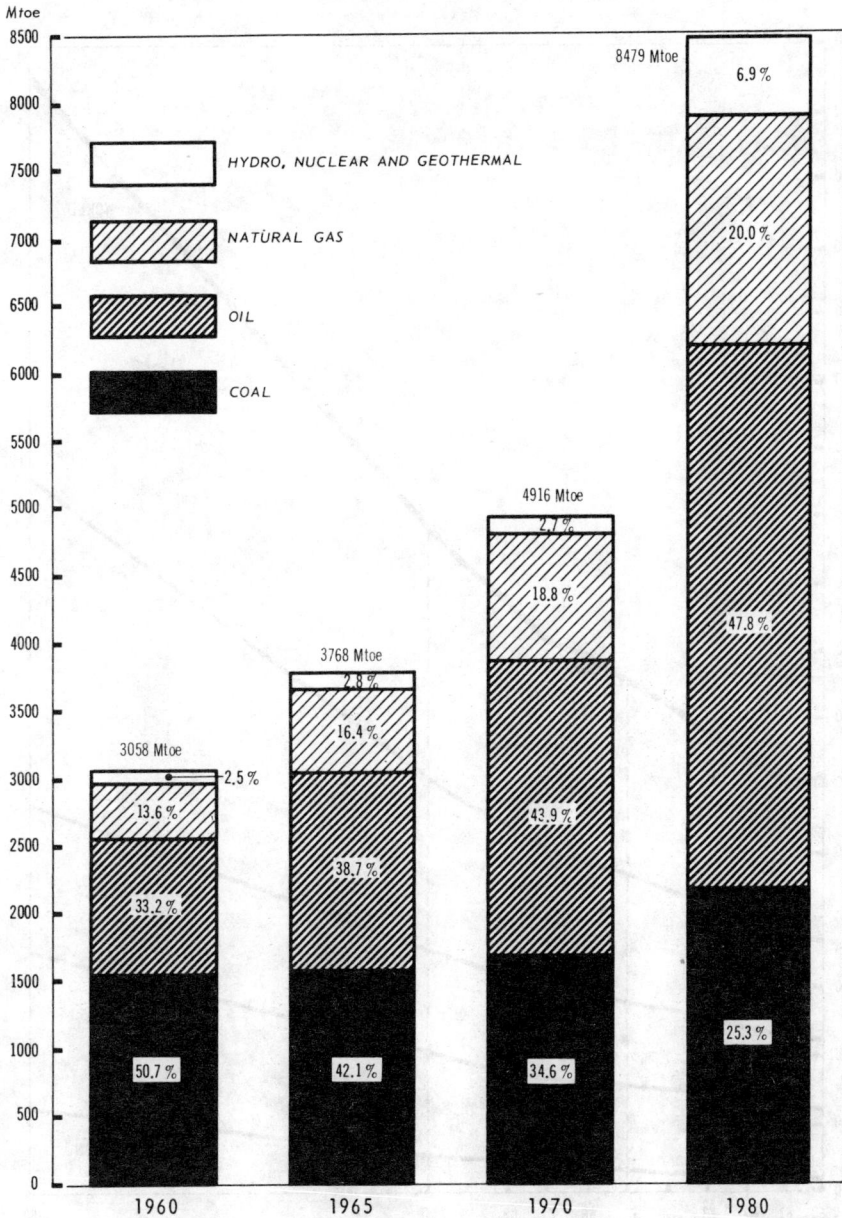

Figure 5

WORLD'S PRIMARY ENERGY REQUIREMENTS

(1960 to 1980)

Million tons of oil equivalent

Figure 6

PRIMARY ENERGY REQUIREMENTS BY AREA
(1960 to 1980)

Million tons of oil equivalent

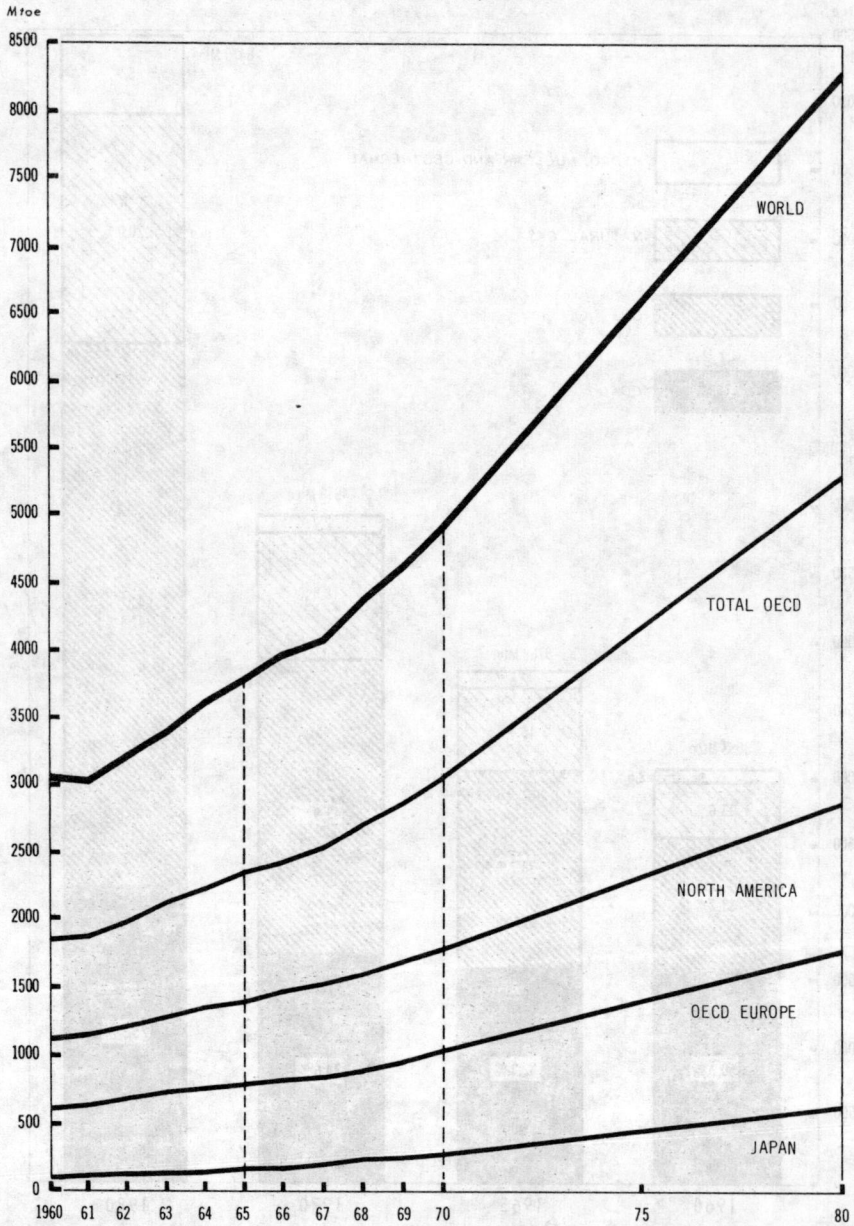

40

Table 1. WORLD'S PRIMARY ENERGY REQUIREMENTS 1960-1980

Million tons of oil equivalent (10^{13} kcal)

	1960		1965		1970		1980		AVERAGE ANNUAL PERCENTAGE GROWTH	
	MILLION TONS	%	MILLION TONS	%	MILLION TONS	%	MILLION TONS	%	60/70	70/80
OECD Europe:	612	100.0	794	100.0	1,041	100.0	1,757	100.0	5.5	5.4
Solid fuels	375	61.3	355	44.7	306	29.4	240	13.7	-0.2	-2.3
Liquid fuels	199	32.6	381	48.0	620	59.6	1,113	63.3	12.0	6.0
Gaseous fuels	11	1.8	19	2.4	70	6.7	231	13.1	20.3	11.9
Hydro, nuclear and geothermal[1]	27	4.3	39	4.9	45.	4.3	173	9.8	5.2	13.9
North America:	1,138	100.0	1,377	100.0	1,763	100.0	2,869	100.0	4.5	5.0
Solid fuels	266	23.4	315	22.9	352	20.0	502	17.5	2.8	3.6
Liquid fuels	511	44.9	600	43.5	763	43.3	1,209	42.2	4.1	4.7
Gaseous fuels	331	29.1	425	30.9	598	33.9	870	30.3	6.1	3.8
Hydro, nuclear and geothermal[1]	30	2.6	37	2.7	50	2.8	288	10.0	5.2	19.4
Japan:	86	100.0	145	100.0	266	100.0	635	100.0	11.9	9.1
Solid fuels	47	54.8	49	34.0	62	23.3	105	16.6	2.7	5.5
Liquid fuels	31	36.4	85	58.6	190	71.7	466	73.4	19.7	9.4
Gaseous fuels	1	1.0	2	1.4	4	1.4	14	2.2	15.2	14.5
Hydro, nuclear and geothermal[1]	7	7.8	9	6.0	10	3.6	50	7.8	3.8	17.7
Rest of world:	1,222	100.0	1,452	100.0	1,846	100.0	3,218	100.0	4.2	5.7
Solid fuels	862	70.5	865	59.6	978	53.0	1,299	40.3	1.3	2.9
Liquid fuels	274	22.4	393	27.1	587	31.8	1,261	39.2	7.9	7.9
Gaseous fuels	74	6.1	173	11.9	251	13.6	582	18.1	13.0	8.8
Hydro, nuclear and geothermal[1]	12	1.0	21	1.4	30	1.6	76	2.4	9.1	9.9
World:	3,058	100.0	3,768	100.0	4,916	100.0	8,479	100.0	4.9	5.6
Solid fuels	1,550	50.7	1,584	42.1	1,698	34.6	2,146	25.3	0.9	2.4
Liquid fuels	1,015	33.2	1,459	38.7	2,160	43.9	4,049	47.8	7.8	6.5
Gaseous fuels	417	13.6	619	16.4	923	18.8	1,697	20.0	8.3	6.2
Hydro, nuclear and geothermal[1]	76	2.5	106	2.8	135	2.7	587	6.9	5.9	15.8

1. Hydro, nuclear and geothermal energy.
SOURCE: For source and other information, see footnotes to Tables, 2, 4, 6 and 8. It should however be noted that some differences in methods and procedures exist between the data for OECD regions and the rest of the world. It follows that the aggregated world figures are indicative only. For OECD areas, the methods and coefficients used in aggregating the different forms of energy are the same as those adopted in the report 'Energy Policy - Problems and Objectives", OECD 1966. See especially Annex II "Sources and Definitions".

Table 2. EUROPE'S PRIMARY ENERGY REQUIREMENTS

1960 - 1980

Million tons of oil equivalent (10^{13} kcal).

	1960	1961	1962	1963	1964	1965	1966	1967	1968	1969	1970	1980[3]
Total requirements	611.8	631.3	677.5	726.3	756.7	793.1	813.7	840.8	899.1	972.0	1 041.4	1 757
Indigenous production[1]	407.5	404.2	411.1	416.9	407.2	402.3	386.8	381.5	395.7	408.5	409.5	749
Coal	330.4	322.6	325.4	325.1	310.4	298.8	278.0	266.7	268.0	266.7	246.5	156
Lignite	23.1	23.5	24.6	25.7	26.5	24.8	23.9	24.1	25.3	26.4	26.5	38
Oil (including NGL)	16.0	17.3	18.4	19.0	21.1	22.2	22.0	21.8	21.8	22.0	23.4	179
Natural gas	11.2	13.0	14.2	14.9	16.3	17.8	20.6	26.1	36.3	49.1	67.7	203
Hydro-power[2]	26.0	26.9	27.1	29.9	29.2	32.7	34.7	34.3	35.2	33.9	35.0	48
Nuclear	0.8	0.9	1.4	2.3	3.7	6.0	7.6	8.5	9.1	10.4	10.4	125
Net Imports[1]	204.3	227.1	266.4	309.4	349.5	390.8	426.9	459.3	503.4	563.5	631.9	1 008
Oil	183.3	206.2	241.0	277.3	318.3	358.7	396.9	430.9	476.6	534.5	596.8	934
Solid fuels	21.1	21.0	25.5	32.2	31.1	31.2	29.0	27.1	25.2	26.5	32.7	46
Natural gas	-0.1	-0.1	-0.1	-0.1	0.1	0.9	1.0	1.3	1.6	2.5	2.4	28

1. Stock changes of coal with indigenous resources, of oil with imports.
2. Includes geothermal energy.
3. Secretariat estimates.

SOURCE: Statistics of Energy OECD 1972.

Table 3.   OECD EUROPE'S OIL PRODUCTS CONSUMPTION PATTERN 1960-1970

YEAR	MOTOR GASOLINE		AVIATION FUELS KEROSENES AND OTHER PRODUCTS[1]		GAS/DIESEL OIL		FUEL OIL		TOTAL ALL PRODUCTS		TOTAL ALL PRODUCTS: PERCENTAGE INCREASE OVER PREVIOUS YEAR
	MILLIONS TONS	%	MILLIONS TONS	%	MILLIONS TONS	%	MILLIONS TONS	%	MILLIONS TONS	%	
1960	29.7	16.2	26.2	14.4	51.3	28.0	75.8	41.4	183.0	100	16.8
1961	33.0	16.0	30.2	14.7	57.8	25.1	85.0	41.2	206.0	100	12.6
1962	36.3	15.1	34.7	14.3	69.0	28.6	101.3	42.0	241.3	100	17.0
1963	40.0	14.5	41.6	15.1	80.7	29.2	113.8	41.2	276.1	100	14.4
1964	44.9	14.3	47.6	15.1	89.4	28.4	132.9	42.2	314.8	100	14.0
1965	49.0	13.9	53.6	15.2	102.5	29.0	148.0	41.9	353.1	100	12.2
1966	53.2	14.1	51.0	13.6	112.4	29.9	159.2	42.4	375.8	100	6.4
1967	57.3	13.8	70.8	17.0	120.8	29.1	166.5	40.1	415.4	100	10.5
1968	62.5	13.4	78.8	16.8	140.6	30.1	185.6	39.7	467.5	100	12.5
1969	67.0	12.8	90.4	17.2	160.7	30.6	206.6	39.4	524.7	100	12.2
1970	72.4	12.4	97.2	16.7	181.5	31.2	230.5	39.7	581.6	100	10.8

1.   Of which Naphtha:

	MILLIONS TONS	% OF TOTAL	% INCREASE OVER PREVIOUS YEAR
1967 ...	17.4	4.2	-
1968 ...	22.3	4.7	28
1969 ...	26.4	5.0	18
1960 ...	27.2	4.7	3

SOURCE: OECD Oil Statistics.

43

Table 4. NORTH AMERICA'S PRIMARY ENERGY REQUIREMENTS

1960 – 1980

Million tons of oil equivalent

	1960	1961	1962	1963	1964	1965	1966	1967	1968	1969	1970	1980[2]
Total requirements	1,137.9	1,152.2	1,206.5	1,258.5	1,317.5	1,378.0	1,458.0	1,504.1	1,599.7	1,683.6	1,762.3	2,869
Indigenous production[1]	1,053.3	1,066.1	1,116.9	1,172.9	1,224.5	1,272.2	1,347.1	1,400.6	1,476.4	1,552.1	1,632.1	2,408
Coal	281.9	276.2	286.0	306.9	321.2	338.0	353.2	347.3	360.0	369.7	389.6	544
Oil	412.4	417.8	433.7	448.8	460.8	473.0	497.2	529.1	554.8	575.8	597.8	731
Of which: NGL	(35.2)	(36.9)	(39.3)	(41.8)	(44.6)	(47.2)	(49.6)	(52.1)	(57.2)	(61.1)	(63.3)	n. a.
Natural gas	329.2	341.4	364.3	384.4	407.2	424.0	458.0	482.0	517.9	558.5	594.9	845
Hydro	29.6	30.1	32.1	31.7	34.1	36.0	36.9	40.0	39.9	44.0	43.8	58
Nuclear	0.2	0.6	0.8	1.1	1.2	1.2	1.8	2.2	3.8	4.1	6.0	230
Net Imports:	84.6	86.1	89.6	85.6	93.0	105.8	110.9	103.5	123.3	131.5	130.2	461
Oil	98.8	100.8	105.6	108.2	114.9	127.3	131.1	124.8	144.9	157.1	164.9	478
Coal	-15.8	-16.0	-18.0	-24.2	-22.8	-22.8	-22.3	-22.6	-22.8	-26.8	-37.7	-42
Natural gas	1.6	1.3	2.0	1.6	0.9	1.3	2.1	1.3	1.2	1.2	3.0	25

1. Stock changes with indigenous production.
2. Co-ordinated estimates submitted by Canadian and United States Delegations. There is some degree of diversity among the experts in estimating the extent to which the United States will be importing oil by 1980. The figures shown reflect the latest Department of the Interior estimates.

SOURCE: Statistics of Energy, OECD.

44

Table 5. NORTH AMERICA'S OIL PRODUCTS CONSUMPTION PATTERN 1961-1970

YEAR	MOTOR GASOLINE MILLIONS TONS	MOTOR GASOLINE %	AVIATION FUELS KEROSENES AND OTHER PRODUCTS[1] MILLIONS TONS	AVIATION FUELS KEROSENES AND OTHER PRODUCTS[1] %	GAS/DIESEL OIL MILLIONS TONS	GAS/DIESEL OIL %	FUEL OIL MILLIONS TONS	FUEL OIL %	TOTAL ALL PRODUCTS MILLIONS TONS	TOTAL ALL PRODUCTS %	TOTAL ALL PRODUCTS: PERCENTAGE INCREASE OVER PREVIOUS YEAR
1961	185.2	39.8	102.1	22.0	103.8	22.3	73.8	15.9	464.9	100	-
1962	192.5	38.2	114.9	22.8	111.2	22.0	85.8	17.0	504.4	100	8.5
1963	198.9	37.7	132.0	25.0	113.6	21.5	83.4	15.8	527.9	100	4.7
1964	206.0	36.4	150.7	26.7	114.8	20.3	93.7	16.6	565.2	100	7.1
1965	211.6	35.6	163.6	27.5	119.3	20.0	100.5	16.9	595.0	100	5.3
1966	221.8	35.4	175.9	28.0	122.7	19.6	106.3	17.0	626.7	100	5.3
1967	228.9	35.6	175.8	27.3	126.3	19.7	111.7	17.4	642.7	100	2.6
1968	243.6	25.9	192.0	28.3	132.9	19.6	110.1	16.2	678.6	100	5.6
1969	255.0	35.0	208.3	28.6	139.5	19.1	125.9	17.3	728.7	100	7.4
1970	267.3	35.1	211.5	27.9	143.6	18.9	137.5	18.1	759.9[a]	100	4.3

1. Of which Naphtha:

	MILLIONS TONS 1967	MILLIONS TONS 1970	ANNUAL PERCENTAGE CHANGE
Special naphtha	2,9	3,6	7,4
Naphtha jet fuel	13,2	10,7	-5,8
Petrochemical feedstock	5,9	6,7	4,4
Total	22,0	21,0	-1,4

(Assuming 8,5 barrels per metric ton.)

2. Of which non-energy:

	1970 THOUSAND B/D	1970 MILLIONS TONS
Lubricants	135	(7)
Asphalt and road oil	440	(26)
Petrochemical feedstocks	818	(40)
Other	201	(9)
Total	1,594	(82)

SOURCE: OECD Oil Statistics.

Table 6. JAPAN'S PRIMARY ENERGY REQUIREMENTS

1960 - 1980

Million tons of oil equivalent (10^{13} kcal)

	1960	1961	1962	1963	1964	1965	1966	1967	1968	1969	1970[2]	1980[2]
Total requirements	86.2	96.6	104.0	118.9	134.3	144.5	163.2	188.7	205.5	237.8	265.7	635.7
Indigenous production[1]	49.2	50.1	49.2	50.9	50.8	48.3	49.9	49.6	47.4	45.3	39.5	82.9
Coal and lignite	41.1	40.6	39.8	40.2	40.3	36.9	37.9	38.7	35.5	33.4	26.3	25.6
Oil	0.5	0.7	0.8	0.8	0.6	0.7	0.8	0.8	0.8	0.8	0.8	2.3
Natural gas	0.9	1.0	1.4	1.9	2.0	2.0	2.1	2.2	2.6	2.6	2.7	5.3
Hydro	6.7	7.8	7.2	8.0	7.9	8.7	8.9	7.7	8.2	8.3	8.6	10.1
Nuclear	-	-	-	-	-	-	0.2	0.2	0.3	0.3	1.1	39.6
Net imports[1]	37.0	46.5	54.8	68.0	83.5	96.2	113.3	139.1	158.1	192.4	226.2	552.8
Oil	30.9	38.0	47.3	59.7	73.8	83.9	98.7	120.1	134.2	162.1	189.6	464.0
Solid fuels	6.1	8.5	7.5	8.3	9.7	12.3	14.6	19.0	23.9	30.3	35.6	79.8
Natural gas	-	-	-	-	-	-	-	-	-	-	1.0	9.0[3]

1. Stock changes of coal with indigenous production, of oil with imports,
2. Data submitted by the Japanese Delegation,
3. Based on firm contracts, Possible new contrats could double the quantities shown (18 Mtoe) - see Chapter IX in which case there could be a corresponding reduction in oil imports,

SOURCE: Statistics of Energy, OECD.

46

Table 7. JAPAN'S OIL PRODUCTS CONSUMPTION PATTERN 1960-1970

YEAR	MOTOR GASOLINE		AVIATION FUELS KEROSENES AND OTHER PRODUCTS[1]		GAS/DIESEL OIL		FUEL OIL		TOTAL ALL PRODUCTS		TOTAL ALL PRODUCTS PERCENTAGE INCREASE OVER PREVIOUS YEAR
	MILLION TONS	%	MILLION TONS	%	MILLION TONS	%	MILLION TONS	%	MILLION TONS	%	
1960	4.0	14.7	2.9	10.7	5.2	19.1	15.1	55.5	27.2	100	-
1961	4.8	14.1	3.8	11.2	5.3	15.6	20.1	59.1	34.0	100	25.0
1962	5.5	14.3	4.6	11.9	6.4	16.6	22.0	57.2	38.5	100	13.2
1963	6.5	12.1	7.7	14.4	7.6	14.2	31.7	59.3	53.5	100	39.0
1964	7.0	9.9	13.2	18.6	8.6	12.2	42.0	59.3	70.8	100	32.3
1965	7.7	9.3	16.7	20.2	9.7	11.7	48.7	58.8	82.8	100	16.9
1966	8.7	9.2	20.4	21.6	11.3	11.9	54.2	57.3	94.6	100	14.2
1967	10.7	9.5	25.4	22.6	12.9	11.5	63.3	56.4	112.3	100	18.7
1968	11.4	8.9	31.0	24.3	14.6	11.5	70.5	55.3	127.5	100	13.5
1969	13.1	8.6	39.3	25.7	16.5	10.8	84.0	54.9	152.9	100	19.9
1070	14.9	8.1	54.6	29.9	19.5	10.7	93.9	51.3	182.9	100	19.6

1. Of which Naphtha:

	MILLION TONS	% OF TOTAL	% INCREASE OVER PREVIOUS YEAR
1967	8.8	7.8	-
1968	11.1	8.7	26
1969	15.3	10.0	38
1970	18.7	10.2	22.2

SOURCE: OECD Oil Statistics.

47

Table 8.   REST OF THE WORLD

196

	1960		1961	1962	1963	1964
Oceania:	31.5	100.0	32.2	34.3	35.7	39.2
Solid fuels	18.9	60.0	18.9	19.6	20.3	21.7
Liquid fuels	11.9	37.8	12.6	13.3	14.0	16.1
Gas	..	..	..	..	..	..
Hydro, nuclear and geothermal[1]	0.7	2.2	0.7	1.4	1.4	1.4
West Asia and Far East:	90.0	100.0	97.8	104.9	113.4	116.4
Solid fuels	45.0	50.0	48.5	52.7	56.9	55.9
Liquid fuels	38.4	42.7	42.1	44.4	47.5	49.9
Gas	5.6	6.2	6.1	6.5	7.4	8.5
Hydro, nuclear and geothermal[1]	1.0	1.1	1.1	1.3	1.6	2.1
Africa:	48.3	100.0	48.3	50.4	53.2	58.1
Solid fuels	30.1	62.4	30.8	30.8	32.2	34.3
Liquid fuels	17.5	36.2	16.8	18.2	19.6	21.7
Gas	..	..	..	0.7	0.7	0.7
Hydro, nuclear and geothermal[1]	0.7	1.4	0.7	0.7	0.7	1.4
South America:	97.3	100.0	102.9	109.9	114.8	121.1
Solid fuels	7.0	7.2	7.0	6.3	7.0	7.7
Liquid fuels	72.1	74.1	76.3	81.9	83.3	85.4
Gas	15.4	15.8	16.8	18.2	20.3	23.8
Hydro, nuclear and geothermal[1]	2.8	2.9	2.8	3.5	4.2	4.2
USSR, other Eastern Europe and China:	954.9	100.0	865.6	908.6	978.8	1,046.5
Solid fuels	760.8	79.7	648.1	658.0	695.0	726.6
Liquid fuels	134.2	14.1	143.7	160.3	176.3	191.5
Gas	52.7	5.5	66.3	81.4	96.2	117.6
Hydro, nuclear and geothermal[1]	7.2	0.7	7.5	8.9	9.3	10.8
Total rest of the World:						
Solid fuels	861.8	70.5	753.3	767.4	811.4	846.2
Liquid fuels	274.1	22.4	291.5	318.1	340.7	364.6
Gas	73.7	6.1	89.2	106.8	126.6	150.6
Hydro, nuclear and geothermal[1]	12.4	1.0	12.8	15.8	17.2	19.9
GRAND TOTAL	1,222.0	100.0	1,146.8	1,208.1	1,295.9	1,381.4

1.   Hydro, nuclear and geothermal energy.

The definitions and method of compilation are those of the United Nations.

The regions are based on the standard United Nations regions except that "West Asia and the Far East" excludes Turkey and Japan; "USSR Other Eastern Europe and China" incl
Yugoslavia. For the full list of countries included under each heading see United Nations Statistical Papers Series J, No. 14 New York 1971. Of the headings above "Oceania"
corresponds to the United Nations Oceania; "West Asia and the Fat East" to the sum under these two United Nations headings; "Africa" to the United Nations Africa; "South
America" to the sum of United Nations Caribbean America and Other America; "USSR Other Eastern Europe and China" to United Nations countries N.E.S.

Million tons of oil equivalent (10^{13} kcal)

1965		1966	1967	1968	1969	1970		1980		ANNUAL AVERAGE PERCENTAGE GROWTH	
										60/70	70/80
41.9	100.0	43.4	45.5	48.3	51.1	54.0	100.0	104	100.0	5.5	6.8
22.3	53.2	22.4	23.1	23.8	24.5	24.3	45.0	39	37.5	2.5	4.8
18.1	43.2	19.6	21.0	23.1	24.5	26.1	48.4	52	50.0	8.2	7.1
..	..	-	-	-	..	1.4	2.6	10	9.6	..	21.7
1.5	3.6	1.4	1.4	1.4	2.1	2.2	4.0	3	2.9	12.1	3.2
128.2	100.0	136.0	144.9	155.5	169.9	187.9	100.0	437	100.0	7.6	8.8
60.9	47.4	61.2	62.1	62.4	64.9	68.0	36.2	188	43.0	4.2	10.7
56.0	43.7	62.9	68.9	76.9	85.4	96.4	51.3	199	45.5	9.6	7.5
9.3	7.3	9.5	11.3	13.2	16.5	20.1	10.7	37	8.5	13.6	6.3
2.0	1.6	2.4	2.6	3.0	3.1	3.4	1.8	13	3.0	13.0	14.3
62.3	100.0	65.1	66.5	69.3	70.7	72.7	100.0	158	100.0	4.2	8.1
37.1	59.6	37.1	37.8	39.2	39.9	40.9	56.2	78	49.4	3.1	6.7
22.4	36.0	25.2	25.9	27.3	28.0	28.9	39.8	61	38.6	5.2	7.8
1.4	2.2	1.4	1.4	1.4	1.4	1.4	2.0	14	8.9	..	25.9
1.4	2.2	1.4	1.4	1.4	1.4	1.4	2.0	5	3.1	7.2	13.6
126.7	100.0	133.7	142.1	155.4	166.6	179.3	100.0	376	100.0	6.3	7.7
7.0	5.5	6.3	7.0	7.0	9.8	9.8	5.5	15	4.0	3.4	4.4
90.3	71.3	93.3	98.7	110.6	116.2	125.4	69.9	249	66.2	5.7	7.1
25.2	19.9	28.7	31.5	32.2	34.3	36.9	20.6	96	25.5	9.1	10.0
4.2	3.3	4.9	4.9	5.6	6.3	7.2	4.0	16	4.3	9.9	8.3
,093.3	100.0	1,160.4	1,125.4	1,220.4	1,285.9	1,355.3	100.0	2,143	100.0	3.6	4.7
737.7	67.4	768.1	699.8	756.6	794.6	822.7	60.7	979	45.7	0.8	1.8
206.2	18.9	225.1	243.3	265.5	291.2	311.7	23.0	700	32.7	8.8	8.4
137.4	12.6	153.4	169.6	184.4	185.0	204.7	15.1	425	19.8	14.5	7.6
12.0	1.1	13.8	12.5	13.9	15.1	16.2	1.2	39	1.8	8.4	9.2
865.0	59.6	895.1	929.8	889.0	933.7	978.1	53.0	1,299	40.3	1.3	2.9
393.0	27.1	426.6	457.8	503.4	545.3	586.8	31.8	1,261	39.2	7.9	7.9
173.3	11.9	193.0	213.8	231.2	237.2	251.0	13.6	582	18.1	13.0	8.8
21.1	1.4	23.9	22.8	25.3	28.0	29.5	1.6	76	2.4	9.1	9.9
,452.4	100.0	1,538.6	1,524.2	1,648.9	1,744.2	1,845.4	100.0	3,218	100.0	4.2	5.7

SOURCES: For 1960-1969 - United Nations, Series J, "World Energy Supplies," for 1970 - Secretariat estimates. The 1960 estimates are indicative only and are based on a variety of sources both internal and external to OECD (see para. 72, footnote 3).

II

THE AVAILABILITY OF OIL

A. TODAY'S PROVED WORLD OIL RESERVES

85.   Table 9 places world "proved reserves" at 79.9 billion tons.   There are, however, no official proved reserves for any country and any world assessment must therefore be treated with caution.   Only the American Petroleum Institute and the Canadian Petroleum Association issue regular assessments of reserves for the United States and Canada respectively; other world figures are gathered by trade journals and there is no consistent rigorous definition of what constitutes "proved reserves".   Moreover, the volume of oil in place even in a well-delineated field can never be precisely accurate; estimates of commercially recoverable oil are usually made not by reference to existing methods of technology but to the production system currently in use, and even this can provide only an approximation.   Assessments of proved reserves have therefore little meaning in terms of absolute world availability; their value is in giving an indication of the location of known oil reserves at a particular point in time and in giving an estimate of the quantity of oil available with present techniques and at current cost.

86.   The ratio of published "proved reserves" to annual production reflects the state of development of each producing area.   This ratio varies for the main areas shown above from 10:1 for the Caribbean to about 70:1 for the Middle East.   As the world average is about 32:1, it appears that statistically sufficient reserves have been delineated to cover requirements for the next 15 years, with consumption increasing at the current rate.

87.   The proved reserve-to-production ratio has remained approximately constant over the last ten years following a ratio as high as 60:1 extant during the preceding five years.   At that time it was relatively easy to discover large oil fields and these reserves related to then current world oil production yielded high reserve-to-production ratios.

88.   From table 9 it is clear that for this decade at least the Middle East will remain the major producing area.   On the basis of present reserves, continuing increases in production may be expected, especially in countries like Saudi Arabia, Iran and Abu Dhabi.

89.   Proved reserves of natural gas, both associated with oil and non-associated, currently represent about 38 billion tons in oil equivalent.   This

51

means a reserve-to-production ratio of about 40:1. Natural gas is likely to feature much more significantly as a source of energy from now on than it has in the recent past; we consider this subject further in chapter IX.

B. OBTAINING RESERVES IN THE FUTURE

90.   Estimates of proved reserves exclude oil resources known to be in place, but not regarded as exploitable using current techniques. As technology develops, more and more of these resources will become available. Moreover, for existing producing fields the average recovery rate is about 30 per cent; an improvement of only 1 per cent would add the equivalent of more than a year's production at current rates. The present scale of exploration gives every reason for expecting further oil discoveries, especially in areas so far little explored, and in particular on the continental shelves, where exploration is in the early stages.

91.   It is clearly impossible to give a precise estimate of the reserves of oil that will come available through advances in recovery techniques. It seems, however, not overly optimistic to put them at the same level as the proven reserves mentioned above.

92.   A major remaining frontier for exploration is to be found on the continental slopes and the continental rises which contain sedimentary basins. Technology is developing rapidly and it seems probable that the deep sea will be a major source of oil in the not too distant future. A deep sea contribution to oil resources, possibly of more than current proven reserves, does not seem too optimistic.

93.   Estimates of ultimate oil reserves are inevitably speculative. They are normally made by comparing, on geological grounds, prospective structures or basins with areas which have been more thoroughly explored and developed. Certain expert opinion has it that the world's onshore and offshore areas contain at least four times the current proved reserves, or more than 300 billion tons.

94.   The estimates above refer to reserves of conventional crude oil, but there are also enormous potential reserves of recoverable or synthetic oils from oil shales, tar sands and coal. While these sources are not yet commercially competitive with existing sources of crude oil, there is no doubt that they will become so in the future and that these ultimately exploitable reserves are far larger than those for conventional crude oil. The amount recoverable from the Athabasca tar sands in Canada alone has been put at roughly equivalent to one-half the total world proved crude oil reserves.

95.   All in all, the amount ultimately recoverable from conventional oil and gas fields, tar sands and oil shales can be considered likely to be perhaps ten times the present total world estimate for proved crude oil and gas reserves, and possibly substantially more. In addition, the oil ultimately available from coal could be even greater than that from all other sources; there is no prospect of a physical shortage of oil for many years ahead.

96.   This is expected also to be true when account is taken of the need to maintain a comfortable proven reserve-to-production ratio. At a yearly

production of 6 billion tons of oil to be expected around 1985, a ratio of 32:1 at the close of 1970 means that proven reserves will have to be brought to 192 billion tons in 1985 from the present level of 80 billion tons. So this means an additional "demand" of about 112 billion tons over a 15-year period.

97.   This calculation is only illustrative; it takes a ratio of 31:1 as fixed. A lowering of the ratio in the future seems, however, acceptable but at a certain point a minimum is reached. The calculation above makes it clear that additions to reserves play an important role in determining the direction of oil technology in the future. It means for instance that the switch from conventional to non-conventional oil may be made earlier than would be expected on grounds of production forecasts only.

98.   We are confident that on purely physical grounds there are ample supplies of oil to sustain the current increasing rate of production for many years ahead. But environmental considerations, especially in North America, may preclude the utilization of oil reserves at what has come to be regarded as a normal pace. Although oil was first discovered on the North Slope of Alaska in 1968, opposition to the building of a Trans-Alaskan pipe-line has effectively denied access to these reserves. Similarly, offshore leasing in the Gulf of Mexico has been delayed by the need to submit environmental impact statements answerable to the National Environmental Policy Act. Moreover, there is growing concern for the conservation of oil resources, to the extent that production levels in certain of the major oil-producing countries have been reduced or at least held constant, rather than allowing increases consonant with past trends.

99.   In this decade most production will continue to come from the present major fields. On present indications Iran and particularly Saudi Arabia and Abu Dhabi are likely to be the main sources of oil entering into international trade. In the Middle East production costs are low but transport costs to the main market areas are high. Certainly there will be new sources (e.g. the North Sea) with higher production costs but lower transport costs and some of these sources will be of significant size. The necessary additions to reserves mentioned earlier will require substantial and expensive efforts in exploring and developing new areas and new sources of oil. In terms of technical costs, it seems that proximity to consuming areas will play a significant role in developing these new areas and new sources. Technical production costs of major oil sources in the deep sea are expected te be in line with the cost of North Sea production, and proximity to consuming areas will again be important. Production costs for oil from tar sands are put at present between $4-4.50 a barrel. This is high for the world as a whole but again proximity to markets is of importance. The wellhead price in the United States for conventional oil of about $3.30 a barrel at present indicates that tar sand oil is not far off commercial large-scale production. The same is true, to a lesser extent, of oil from shale. Vast reserves lie in the United States and although production costs of shale oil will be in the neighborhood of $4 to $5 a barrel, depending upon assumed return on investment, the amount of production and the like, the first commercial production may be expected here. Reserves of oil shales other than tar sands are spread widely throughout the world and outside the United

States the shales which are closest to the market will come first into production. Just when these sources will become commercial is difficult to say, but it is clear that the necessary investment for exploration and production in these new areas and forms will be a multiple of current investment. This more general question is examined in chapter VIII.

Table 9. PROVED WORLD OIL RESERVES AT END 1971

	BILLION Bbls.	BILLION TONS[1]
United States	45.4[2]	6.2[2]
Canada	10.2[2]	1.4[2]
Caribbean	17.1	2.3
Other Western Hemisphere	14.5	2.0
Total Western Hemisphere	87.2	11.9
Middle East	367.4	50.3
OECD Europe	13.9	1.9
USSR, Eastern Europe and Chinese People's Republic	40.2	5.5
Africa	58.9	8.1
Other Eastern Hemisphere	15.9	2.2
Total Eastern Hemisphere	496.3	68.0
Total world	583.5	79.9

1. Converted at rate of 7.3 barrels per ton.
2. The figures for the United States and Canada include natural gas liquids.
SOURCES: United States: American Petroleum Institute.
        Canada: Canadian Petroleum Association.
        USSR, Eastern Europe and China: US estimate.
        Other Areas: Oil and Gas Journal (issue of 27th December, 1971).

III

SOURCES OF OIL SUPPLIES

A. WORLD CRUDE OIL PRODUCTION 1965-1970

100. Table 10 illustrates changes in the pattern of world production of crude oil during the period 1966 to 1970. All the trends noted in the 1964 Report[1] have been at least maintained, if not accentuated. Output from Libya and Nigeria, reflecting proximity to the main consuming markets, exhibited the highest relative growth rates among the several major producing areas of the world. While in 1962 Africa's contribution was three per cent of total world crude oil supplies, by the close of the decade this had risen to 13 per cent and thus had provided a quarter of the total world increase in crude oil production between the years 1965 and 1970. The actual increases were:

Producing Area	Million tons	Per cent
World	783.2	100
United States	102.0	13.0
Middle East	273.6	34.9
Africa	192.6	24.6
USSR, other Eastern Europe and China	121.8	15.6
Others	93.2	11.9

101. The United States position illustrates again the difficulties of expanding upon a crude oil producing base of advanced maturity; a relative decline among the important oil producing areas of the world continued despite advances in absolute terms. Moreover, North America during these years has been overtaken, in terms of proportional contribution to world production, by the Middle East; indeed imports into the United States, at least in the latter years of the 1960s, have risen steadily to fill the growing gap between demand and indigenous supply.

102. The Middle East is now the dominant crude oil producing area of the world; the size of its reserves in comparison with those other areas makes it very likely that this preeminence will be preserved for the remainder of the present decade. The Middle East now produces more crude oil than

---

1. "Oil Today - 1964", OECD, 1964.

any other area— 29 per cent of the world total in 1970— and as noted above, it accounted by far for the largest part— 35 per cent— of the world gain in production during 1966-1970.

B.   THE SOURCES OF OECD EUROPE'S OIL SUPPLIES

103.   As table 11 shows, a major change has taken place in the geographical origin of OECD Europe's oil supplies between the years 1960 and 1970.   The most striking development has been the increase in the relative share of imports from Africa, from 5 per cent in 1960 to more than one-third in 1970.   In parallel herewith, the Middle East saw its relative share reduced from 73 per cent to 50 per cent; in actual volume the increase from the Middle East has of course been very considerable, total exports to Europe having more than doubled during the ten years.   A further change to be noted was the relative decline in the importance of Venezuela as a source of supply.

104.   Europe's own crude oil production contributed less than 4 per cent of its requirements in 1970, notwithstanding a sustained level of exploration efforts over the years.   Our last report mentioned a major exploration campaign which was then being launched on the continental shelf in the North Sea as a consequence of the gas discoveries in the Netherlands. Since then major gas and oil discoveries have been made in that area and we believe that oil supplies from the North Sea may contribute 13 per cent or more of European oil demand by 1980.

105.   Though indigenous supplies of that order are a welcome contribution, Europe, in comparison with the United States, will still have to import by far the greatest part of its requirements from overseas sources, located mainly in areas which in recent years have been subject to continuing unrest.

106.   Over the period up to 1980, it is likely that the Middle East's share of supplies will regain some of the position lost during the past decade, up to perhaps 55 to 60 per cent, which implies that the area will continue to dominate the European supply position, as it has done until now.

PRODUCTION IN OECD EUROPE

107.   In relation to total requirements, production from European fields has been of diminishing importance; their share has decreased from some 7.9 per cent in 1960 to less than 4 per cent in 1970.   The major exploratory effort is now concentrated on the Continental Shelf, but no discoveries of outstanding importance have been made.

108.   Oilfields on the continent produce high cost oil in general, the advantage of short distance to consuming areas being offset by production costs.   Security of supply is a positive factor, however, and several countries have adopted measures to encourage indigenous oil production for that reason.

109.   The centre of activities of oil exploration in OECD Europe during the last several years has been around the drilling platforms in the North

Sea. In 1970 the outlook for indigenous oil supplies improved substantially as a result of the discovery in the North Sea of oil in significant quantities. The discovery was of prime importance; the existence of major accumulations of oil in this part of the world was substantiated.

110.   At the present stage it is hazardous to make precise forecasts on the development of North Sea production. On the basis of certain preliminary assessments with regard to the discoveries so far announced, it is likely that production, which started at a modest rate in 1971, could reach 50 million tons in 1975 and approach 150 million tons by 1980. Including production on land, indigenous West European oil production may therefore reach 16 per cent of European oil requirements by 1980.

IMPORTS

111.   While taking into account the North Sea discoveries, Europe will still have to rely during the seventies for the majority of its requirements on outside sources, from 96 per cent in 1970, declining to 85 per cent by 1980. These sources are located mainly in three geographical areas: a number of producing states surrounding the Persian Gulf and commonly referred to as the Middle East; the North African area, including Libya, Algeria and Egypt; and thirdly, Nigeria or more generally West Africa.

112.   The proportionate amounts of imports from the respective areas will depend on a number of factors, such as crude quality, cost, distance of transport, public policies and corporate considerations.

113.   Variations in quality of available crude oils are considerable, leading to corresponding differences in product yield and in sulphur content, to mention the more important factors. Apart from considerations of cost at source and distance, the very rapid increase of imports of light African crude oils, mainly from Libya, into Europe can be partly explained in terms of quality. The greater yield of light fractions, in particular middle distillates, and the lower sulphur content of African crude oils were more in line with the demand pattern in Northwestern Europe in particular.

114.   The geographic pattern of sources of crude oil supply has changed considerably over the years 1960-1970. As has been noted, the outstanding feature of this change was the upsurge of imports from North Africa, increasing from 8 million tons in 1960 to 194 million tons in 1970. The extremely great dependence on one source of short haul crude oil by the end of the past decade was a singularly striking feature of that period.

115.   The share of supplies from the Middle East decreased from some 73 per cent to 50 per cent in the past ten-year period, mainly as a result of its geographical disadvantage in relation to European markets, in particular since the closure of the Suez Canal in June 1967, but also because of the individual companies' availability pattern. Venezuelan crude is high cost. Because the period under review was characterized by heavy competition from low cost crude oil sources and an erosion of product prices, supplies from Venezuela lost ground. Supplies from this source continue to be imported largely out of the necessity to balance variations in the types of crude oil from other sources.

116.   Supplies from the Middle East are likely to remain the mainstay of OECD Europe's requirements over the next decade and may cover some 60 per cent by 1980.   The bulk of the world's known oil reserves outside the United States and the USSR are concentrated in the countries surrounding the Persian Gulf.   A small number of extremely large fields supply the greater part of present production.   The possibilities of discovering new fields are certainly not exhausted; witness the development of recent finds in Abu Dhabi, Oman and Dubai.

117.   Production in the Middle East may well double during the 1970s and the share of the area in world production may increase from some 30 per cent in 1970 to over 35 per cent in 1980.   If demand should exceed the forecasts based on present assumptions, most of the additional requirements are likely to be supplied from the Persian Gulf producing areas.

118.   Production in Africa as a whole went up from 12 million tons in 1960 to 300 million tons in 1970, the major part of the increase being accounted for by Libya, where in particular a number of American companies, new to the area, have rapidly developed a number of fields. It seems likely that over the period up to 1980 the increase in North African production may be at a slower rate, based on current trends in the area. Production controls purportedly for conservation purposes were introduced in Libya in the course of 1970.   Algerian production has suffered from an interruption of loadings during the course of 1971.   Production in West Africa, on the other hand, and particularly in Nigeria, has been developing rapidly after the cessation of internal hostilities in 1969.   Production from that area may well approach 200 million tons by 1980, as compared to a level of 65 million tons in 1970, ranking the area almost equally with North Africa as a crude oil producer by the end of the decade.

119.   Substantial reserves of oil exist in the West Siberian regions of the USSR, but their development will be hampered by technological and transport problems.   Their location in relation to consuming areas within the USSR and other Eastern European countries requires major investments in transport.   Production is therefore unlikely to increase to such an extent that a major increase in exports to Western Europe could be expected.

120.   Currently the trade position of Eastern Europe at present shows net exports of some 52 million tons per year.   It is unlikely that the volume of exports from the area into Western Europe will grow substantially over the period under review, and they will therefore decline in relative terms. Moreover, because the import of oil by Eastern Europe in the coming years is likely to grow faster than exports, a weakening of the net trade position can be expected.

121.   The closure of the Suez Canal to all shipping in June 1967 as a result of hostilities in the area, led to the necessity of re-organizing supply and shipping arrangements for Western Europe at very short notice, and in such a way that no shortfall should occur in any of the consuming areas. Almost overnight the bulk of Middle East oil had to be carried via the much longer shipping route around Africa.   Though the disruption of normal routes did not lead to any physical supply shortage, a considerable

increase in cost of transport was unavoidable, a reflection of the longer distance to be covered and the rise in freight rates. When the supply of tankers had adapted itself to new demand levels, the rates came down to more normal levels and eased the position of the non-integrated supplier, but the effect of distance on delivered cost in Europe could only be gradually reduced by the introduction of very large crude carriers in the 200,000-250,000 dwt range.

122.   The building of new pipe-lines for carrying oil from the Middle East to the Mediterranean, may render the geographical transport pattern less flexible and would present a serious risk of interruption in case of an emergency of any kind.

123.   Assuming that the projects for pipe-line construction in this area, which are known and firm, would be carried through, some 15 per cent of Europe's supplies would be delivered via Eastern Mediterranean pipe-lines by 1980, as against 10 per cent in 1970. For the other areas the approximate percentage distribution of origins by 1980 could be of the following order: Persian Gulf 45 per cent (via the Cape), North Africa 15 per cent, and West Africa 10 per cent, the remaining 15 per cent to be made up by indigenous production.

124.   The cost of crude oil delivered at European points of destination has risen considerably over the past five years, the major part of the increase being accounted for by the effects of the agreements reached between oil companies and producing countries early in 1971. The increase has also reflected rises in transport costs since June 1967.

125.   The second and more far-reaching growth in the cost of supply came with the conclusion early in 1971 of the agreements regarding new fiscal conditions, reached between the governments of the producing countries in the Middle East and Africa and representatives negotiating on behalf of all oil companies with producing interests in these countries. The outcome of the negotiations was of particular importance to Western Europe and Japan, which are almost entirely supplied from these areas.

126.   Under the terms of the agreements, the basis for assessing companies' tax liabilities— posted prices— was increased in particular in Libya and for the Eastern Mediterranean points of loading, and the rate of taxation was brought up from 50 to 55 per cent. The cumulative effect was a very substantial rise in the tax-paid cost of crude oil at seaboard both at Persian Gulf and Mediterranean points of loading. For a light Persian Gulf crude oil the rise was about 40 cents per barrel, against some 90 cents per barrel for crude oil from Libya. The delivered tax-paid cost in Northwest Europe for an integrated supplier had risen by some 25 per cent for a light Persian Gulf crude and by 50 per cent for a Libyan crude oil, at normal freight levels and the cif price for Northwest Europe reached about $2.80-$3.00 a barrel.

C. THE SOURCES OF NORTH AMERICA'S OIL SUPPLIES

127.   In terms of security of supplies, North America today is in a much more favoured position than OECD Europe and Japan, simply because

North American indigenous production is sufficient to cover almost four-fifths of requirements. The sources of North American oil supplies are illustrated in table 12.

PRODUCTION

128.   A comparison of rates of growth in the production and consumption of crude oil and petroleum products in North America as a whole, with those rates for Europe and Japan, can be somewhat misleading, inasmuch as the United States, the dominant partner, is proceeding from a much larger base, and from a more mature economy which cannot be expected to generate comparable growth rates. To illustrate, United States production increased by just 17.6 per cent between 1965 and 1970, while Canadian production, less than an eighth of the US level, rose by 57 per cent. Yet the absolute increment obtained in the United States was three times that secured by Canada.

129.   The geographic pattern of crude oil production in the United States and Canada has held relatively constant in the years past, and no major change is anticipated until the Trans-Alaskan crude oil pipe-line is completed and the extraction of crude oil on the North Slope of Alaska becomes a reality. Similarly, Canada is looking to its Arctic as well. In the interim, the production of crude oil will remain concentrated in the central and western areas in both countries, far removed from the major consuming centers in the east. Canada has found it useful to import quantities of oil into its eastern province as a means of covering local requirements rather than extending the transcontinental pipe-line to link buyer and seller; delivered costs of foreign oil still carry an economic advantage.

130.   The United States has continued its search for crude oil and gas into offshore waters of the Gulf of Mexico and more recently in the Santa Barbara Channel of California. This search, although expensive in that exploration and production from the deeper waters demand an entirely new array of techniques and equipment, has been fruitful, and the oil and gas coming from the fields which have been discovered are important contributors to total supplies.

IMPORTS

131.   The United States, after enjoying many years of cheap and plentiful supplies of energy, is entering a period where increasing demand for energy is paralleled by growing problems of supply. These problems, while still local in scope and seasonal in nature, are the product of progressive declines in proved reserves of oil and gas and in crude productive capacity; and of restrictions imposed on the extraction and utilization of energy, brought about by environmental considerations; they are in part attributable to the inflexibility associated with growing imports of energy.

132.   Recognizing that the task of providing energy, in sufficient amounts and in useable form, to the economy in the coming years is further complicated by the long lead times required to expand supplies, the United States has taken, or is contemplating, actions which will broaden its energy base, in accordance with the President's clean energy message of June, 1971.

A number of steps have been taken which will lead to the orderly development of oil shale resources, with competitive sales of lease tracts to begin in late 1972. A tentative schedule for oil and gas leasing on the Outer Continental Shelf[1] has been developed for the period through 1975, with the Gulf of Mexico, the Gulf of Alaska and the Atlantic Coast covered. Additional funds have been allocated to pursue a coal gasification programme. Positive action is being taken with regard to the possible leasing and development of geothermal resources. Means for the removal of pollutants from stack gases or, alternatively, reduction in the sulphur content of coal to acceptable levels prior to burning, are being actively sought in order that the contribution of coal to energy supplies can be more in keeping with its availability.

133.   Finally, the potential of oil and gas resources in the United States exceeds by far the performance of the domestic oil and gas industry. It has been estimated that the amount of crude oil yet to be discovered in the United States is equal to all of the oil found so far. To realize this potential, however, will require the proper interplay of economic, political and technological factors.

134.   When the United States imposed mandatory oil import controls, such action was taken for the express purpose of preserving national security, but there were also a number of related objectives, including the encouragement of discovery and development of domestic petroleum reserves. Import controls have been used successfully to assure a market for domestic production and to lend a measure of stability deemed necessary to the rational planning of future operations. Nevertheless, a number of changes and exceptions have been made over the years which in sum now make for an oil import programme quite unlike that originally imposed. Perhaps most important, restrictions on the import of residual fuel-oil have been removed which has led to a very rapid rise in the imports of this product, especially for the East Coast area, where such imports reached 75 million tons or 90 per cent of East Coast supply in 1970. For 1972, and expanding on actions taken in the preceding two years, limitations on the import of crude oil are being further relaxed. Finally, the emergence of new domestic interests, seeking imports of selected raw materials as a means for preserving their competitive position, have brought other changes to the programme. In essence, the original import programme is being remodelled to permit reasonable response to the need for additional energy, at prices favourable to the consumer, and in volumes consistent with national security considerations.

135.   Despite the tremendous potential for oil and gas yet untapped in the United States, it is likely that imported oil will be looked upon as the principal

---

1. It is important to understand the distinction between State-owned offshore lands and Federal offshore lands. The Outer Continental Shelf comprises that part of the Continental Shelf that is under Federal jurisdiction. Along the Atlantic and Pacific coasts the Outer Continental Shelf begins at a line three geographical miles from the coastline. In the Gulf of Mexico it also begins 3 geographical miles from the coastline, except offshore Texas and Louisiana it begins 9 geographical miles from the coastline of those States. Seaward of these markings the Federal Government assumes jurisdiction and control, and the granting of mineral leases is the responsibility of the Department of the Interior.

supplementary source of supply for the remainder of this decade. In 1970 imports provided 22 per cent of the demand for petroleum (crude oil and petroleum products) in the United States. It has been estimated that by 1980 the United States, with Alaskan North Slope production projected to a level of no more than 75 million tons, would need some 480 million tons of oil from Western and Eastern Hemisphere sources to balance supply and demand. In that year, then, imports would account for 43 per cent of demand; any reduction in the indicated availability of North Slope oil would have to be offset by increased imports and would further enlarge this dependence.

D. THE SOURCES OF JAPAN'S OIL SUPPLIES

136.   Due to the meagreness of the indigenous crude oil production, the rapid increase in demand for oil in Japan has, of necessity, always been paralleled by a growth in oil imports. The following paragraphs describe the development in geographic origin of Japanese oil imports since 1955 (see also table 13). The main feature has been the dominance of the Middle East as a supplier of oil.

137.   In 1955 the amount of crude oil imported by Japan was as little as 7.4 million tons. In the 15 years that followed, imports were increased by more than 22 times, amounting to nearly 170 million tons in 1970.

138.   In 1955 Japan imported from the Middle East 5.6 million tons of crude oil, that is to say, about three-fourths of its total crude oil imports during that year, and from the Far East— 1.8 million tons— or about 24 per cent, while the United States supplied the remaining 1 per cent.

139.   In 1960, the Middle East supplied Japan with 21.5 million tons, or about 80 per cent of her total crude oil imports. The Far East came next with the figure of 4.3 million tons, a share of 16 per cent. Eastern Europe, which emerged as one of Japan's sources of crude oil in 1958, exported about 1 million tons to Japan in 1960, equal to 4 per cent of its total crude oil imports.

140.   In 1965, the share of supplies from the Middle East among the total imports of crude oil by Japan reached approximately 88 per cent, the amount of crude oil imports from that area being 63.8 million tons. Correspondingly, the share of supplies from the Far East dropped to 7 per cent, or to 5.4 million tons, or 4 per cent of the imports.

141.   In 1970, the Middle East remained the most important source of oil supplies for Japan, which imported 144.8 million tons from this area. In relative terms, the Middle East declined, if only slightly, registering 85 per cent of total crude imports. Japan's oil imports from the Far East that year were 22.6 million tons, or 13 per cent.

142.   Since 1955 the Middle East has been by far the largest supplier of oil to Japan, with the Far East a distant second. Other sources have been of only minor importance. Until the late 1960s Japan's demand for oil was doubling every three or four years. Even so, the participation of the Middle East in the Japanese oil supply was enlarging even more rapidly.

Iran contributed most to this tendency, while Kuwait levelled off in its exports to Japan in the latter half of the 1960s, and Saudi Arabia has demonstrated a somewhat similar trend recently.

143.   The Far East, largely Indonesia, had not been able to take advantage of the expanding Japanese market and its relative share declined to just 6 per cent in 1967.   But in the following years, recovery was substantial, to the extent that 13 per cent of the market had been captured by 1970.   The resurgence of the Far East as a source of oil for Japan may be mainly attributed to a marked demand in recent years for low-sulphur crude oil.

144.   Other sources of supplies have never played any particularly important role in imports into Japan, with the exception of Eastern Europe that at one time supplied 5 per cent of Japan's total crude imports.   Africa recently initiated exports to Japan, while the Western Hemisphere has long been a supplier.   But the amounts involved are very small compared with those of the total imports of crude oil by Japan.

145.   As is shown in table 6 of chapter I, oil imports will increase year by year, reflecting the growth in energy demand.   One estimate for oil imports in 1980 amounts to 460 million tons.   Most of this import will take the form of crude oil.   And, in 1980, crude oil imported into Japan will mostly be from the Persian Gulf countries which are presently the main source of supply for Japan, and from Indonesia.   A diversification of the sources of supply may be realized with increased imports from African countries, the USSR, etc., if their oil is competitive.

146.   Prospection in the continental shelves around Japan is actively taking place and in the event of oil fields of a substantial scale being discovered, the amount of future oil imports will accordingly be different from that which is anticipated today.

E.   THE SOURCES OF AUSTRALIA'S OIL SUPPLIES

147.   Australia's first commercial oil field was discovered at Moonie in Queensland in 1961.   Production started in 1964 in conjunction with three small adjacent fields.   Subsequently, other commercial fields have been brought into production at Barrow Island, Western Australia, in 1967 and in Bass Strait, offshore from Victoria, in 1969.

148.   Until 1964 Australia was wholly dependent on imports for the total of its oil supply and even in 1969 still imported 94 per cent of all its requirements.   These supplies were traditionally obtained about 75 per cent from the Middle East and 25 per cent from Far East sources.   See origin of oil supplies in 1960, 1965 and 1970 in table 14.

PRODUCTION

149.   Over the last 15 years the search for oil in Australia, aided by Commonwealth Government measures, has been widespread.   Drilling has been more successful in the offshore areas.   About 80 per cent of proven

crude oil reserves are located in the Bass Strait fields. These deposits are close to the large markets in Sydney and Melbourne.

150.   Under the Government's oil policies, Australian refineries are required to purchase and refine all locally produced crude oil up to a level set by an absorption formula. There is a limit to the amount which can be absorbed in that Australian crudes found so far are of a light nature resulting in a shortfall of those fractions from which bitumen, lubricating oil and fuel-oil are produced.

151.   To correct this imbalance, it is necessary to import heavy crudes or other feedstocks for manufacturing bitumen and lubricating oil, along with a part of fuel-oil requirements. The remainder of the fuel-oil demand is directly imported. With the likely pattern of production from known fields, and the forecast usage of different products, the maximum absorption rate possible to 1980 will be approximately 67 per cent of total demand. Production from existing fields is now close to this proportion of the present demand, but unless substantial new discoveries are made in the near future, production is expected to fall well below this limit by 1980.

152.   There are a number of recently discovered hydrocarbon deposits. These have not yet been fully evaluated, but appear at this stage to be mainly of natural gas rather than crude oil.

Table 10.   WORLD CRUDE OIL PRODUCTION 1960 AND 1970[1]

COUNTRY/AREA	MILLION TONS		PERCENTAGE OF WORLD TOTAL		PERCENTAGE INCREASE 1966-1970	PERCENTAGE AVERAGE ANNUAL INCREASE, 1966-1970
	1965	1970	1965	1970		
United States	436.0	538.0	27.8	22.8	23.4	4.4
Canada	45.0	67.6	2.9	2.9	50.2	8.5
Other Western Hemisphere	242.7	276.4	15.4	11.8	13.9	2.6
Middle East	415.5	689.1	26.5	29.3	65.8	10.6
Africa	107.3	299.9	6.8	12.7	179.5	22.8
Western Europe	22.1	22.9	1.4	1.0	3.6	0.7
USSR, other Eastern Europe and China	268.2	390.0	17.1	16.6	45.4	7.8
Other Eastern Hemisphere	32.8	68.9	2.1	2.9	110.1	16.0
World Total	1,569.6	2,352.8	100.0	100.0	49.9	8.5

1.   Including natural gas liquids and shale oil.

67

Table 11. OECD EUROPE CRUDE OIL SUPPLY
1960, 1965, 1970

Million tons

	1960	1965	1970
Production			
Crude oil	14.5	20	21.8
Natural Gas Liquids	-	0.7	0.7
Total production	14.5	20.7	22.5
Imports crude oil[1]			
Western Hemisphere	17.8	24.9	24.2
Middle East	133.7	...9 3	309.0
North Africa	8.1	7..	193.7
West Africa	1.0	14.6	42.7
Other	8.4	13.2	24.2
Total Crude Imports	169.0	328.2	593.8
Total supplies	183.5	348.9	616.3
Imports as per cent of supply	92.1	94.1	96.3

1. Excluding movements in the European area.
SOURCE: OECD Oil Statistics.

Table 12. OECD NORTH AMERICA OIL SUPPLY
1960, 1965 AND 1970

Million tons

	1960	1965	1970
Production			
Crude oil (including shale oil)	358.8	408.4	517.0
Natural gas and liquids	33.7	47.4	67.3
Total production	392.5	455.8	584.3
Imports of crude oil[1]			
Western Hemisphere	35.1	35.8	37.9
Middle East	22.1	23.4	16.1
Other	3.8	6.4	12.8
Total crude imports	61.0	65.6	66.8
Imports of petroleum products[1]			
Western Hemisphere	39.3	55.5	89.2
Middle East	1.2	0.7	1.8
Other	0.4	0.5	8.7
Total product imports	40.9	56.7	99.7
Total imports	101.9	122.3	166.5
Total consumption	512.8	594.0	755.8
Total imports as a percent of total consumption .	19.8	20.5	22.0

1. Exclude movements between United States and Canada.

Table 13.   JAPAN'S CRUDE OIL SUPPLY
1955, 1960, 1965 AND 1970

Million tons

	1955	1960	1965	1970
Production				
Crude oil	0.3	0.5	0.6	0.8
Imports of Crude oil				
Middle East	5.5	21.5	63.5	144.5
Far East	1.7	4.1	5.3	22.5
Others	0.1	1.3	3.4	2.5
Total imports	7.3	26.9	72.2	169.5
Total Supply	7.6	27.4	72.8	170.3
Imports as percent of supply ..	96.1	98.2	99.2	99.5

Table 14. AUSTRALIA'S OIL SUPPLY
1960, 1965 AND 1970

Million tons

	1960	1965	1970
Crude and process oils			
Local production	-	0.3	8.1
Imports			
Middle East	7.5	11.7	12.4
Far East	3.6	4.8	3.5
Other	0.2	-	-
Total supply	11.3	16.8	24.0
Refinery Intake	11.3	16.7	23.8
Exports	-	-	0.1
Total Disposals	11.3	16.7	23.9
Petroleum products			
Refinery output	11.2	14.7	21.1
Imports			
Middle East	0.5	0.6	1.7
Far East	0.6	0.5	0.6
Other	0.1	0.5	0.3
Total supply	12.4	16.3	23.7
Deliveries to Inland Consumption	9.5	13.3	19.9
Bunkers	1.4	1.7	2.3
Exports	1.6	1.1	1.4
Total Disposals	12.5	16.1	23.6

IV

THE SUPPLY OF OIL

A. THE STATE OF THE OIL MARKET

153.   During the past decade the demand for hydrocarbon fuels of all kinds in all countries has exceeded even the most optimistic forecasts. Rapid industrial development and economic expansion have been accepted as normal and have drawn on energy supplies secured at the lowest rates.

154.   The low level of prices at which oil (and gas) was available during the 1960s further stimulated the use of these fuels.   The level largely reflected the emergence of new producing areas in Africa and of state-owned oil companies in many of the host nations.   The pressure of the newcomers on the crude oil market was matched by the efforts of the established companies to increase or at least to maintain their shares of the petroleum product market.   The inevitable depression in price levels meant that in some areas the return on invested capital was very low.

155.   Despite the burgeoning demand for oil worldwide, supplies in this period have always been ample; there have been no instances of industrial slow-downs or consumer discomfort attributable to shortages of oil.   The experience of the Suez crisis in the mid 1950s helped the industry and the consumer alike to weather the Middle East political emergency of 1967.   The Six Day War led to the closure of the Suez Canal, a short-time interruption of supplies and a selective embargo on supplies to certain consuming countries.   The resulting pressure on the tanker market was contained through co-operation within the oil industry.   The International Industry Advisory Body (IIAB) assisted the OECD Oil Committee in assessing the oil supply and transportation position throughout the crisis.

156.   Moreover, the gathering momentum of a number of changes in factors basic to the international oil industry coincided in drastic alterations in the state of the oil market in 1970.

157.   Demand for energy, increasing at a rate greater than the increase in gross national product, exceeded forecast levels.   Coal was beset with a variety of problems, both in Europe and in the United States, and the construction of nuclear electricity generating facilities fell far short of expectations.   The continued closure of the Suez Canal, the reduction in Libyan oil supplies, the interruption of TAPLINE service and the initial shortage of tanker capacity combined to bring about a shortage of crude oil

71

for Europe. In consequence, what for most of the decade had been a buyer's market was transformed quite rapidly into a seller's market.

158. There can be no doubt that the state of the oil market as it entered the decade of the 1970s was completely unlike that market which existed at the beginning of the 1960s. The change to a seller's market brought with it the inevitable change in consumer prices. Moreover, it enabled the producing nations and the national oil companies of these nations to seek and secure for themselves greater control over the market. As a further consequence, the relation between the international oil company, the host nation and the consumer were substantially altered. Even so, oil markets can be volatile: they have the characteristic of rapid response to economic or political developments which might affect levels of supplies or transport facilities. Indeed, the decline in tanker rates during the latter part of 1971 demonstrated that the seller's market was not immune to change. The drop in Libyan production reflected the loss of at least part of the short-haul attraction of those supplies as the influence of a rapidly expanding fleet of very large crude carriers— and a decline in combined dry cargo charters— began to be felt. Moreover, the falling-off of economic growth in certain oil markets showed that the seller's market itself was now shifting, if only slightly, in favour of the consumer.

159. Another important change in the latter years of the past decade was that exploration and development of new petroleum reserves within the United States have no longer kept pace with demand. Consequently, shut-in production capacity, looked to as a strategic emergency reserve to be drawn upon in the event of interruption of normal supply routes, has now disappeared for all practical purposes. Indeed, since about 1967, the United States itself would have been unable to cover all of its domestic oil requirement should foreign oil be unavailable.

160. Both consumer and producer governments have either embraced new strategies or expanded upon existing policies which ultimately influence the pattern of oil movements. The need to provide for secure sources of energy has in varying degree brought about in some consuming areas the enforced protection of indigenous fuel industries and, as well, the strict control of relative contributions to total supply by the competing fuel materials. Inflation and the seemingly insatiable appetite for investment funds in all countries have been reflected in a growing awareness of cost and price trends, the desire to protect domestic markets against foreign incursion, the importance of future revenue expansion and the desirability of maintaining the balance of payments under reasonable control. All these factors point to need for stability in the international oil market, stability which extends to the uninterrupted movement of oil to the consumer, in the desired amounts, and to the posted prices which determine revenue for the producer.

161. Eastern Europe, in its continued effort to expand sales of oil, and to maximize hard currency earnings from such sales continued to concentrate on markets in Western Europe. Although these sales increased from year to year, growth was less than had been forecast by some observers. This, in conjunction with high rates of increase in demand in these markets, meant that the importance of East European oil in total European supply was less

at the end of the decade than at the beginning. Continuing need for foreign exchange, or the means to barter manufactured goods and advanced technology, found expression in contracts for Soviet natural gas exports to selected OECD European countries. These long-term deliveries, scheduled to begin by the early to mid-1970s, will be made under reciprocal trade agreements.

162. The easy conditions in tanker transport facilities, as an adjunct to the substantial improvement in the tanker fleet following the Suez crisis of 1956-57, prevailed until the closure of the Suez Canal in June 1967. Indeed, at the time of the closure, the world tanker fleet had been in considerable surplus to current needs. Tankers were being operated at slow speeds to reduce costs; some were being routed round the Cape of Good Hope or loading in the eastern Mediterranean or passing through the Suez Canal. The denial of the Canal as an aftermath of armed conflict in the Middle East in 1967 and the failure to find mutually acceptable terms under which this artery might be reopened have led to a dramatic shift in tanker movements out of the Persian Gulf. The bulk of the oil now moving out of the Gulf is directed around the Cape of Good Hope. The use of this route to Europe, adding some 25 days to the round trip via the Suez Canal, in addition to increasing the demand for tanker capacity, helped to precipitate a change in the pattern of tanker use. New, very large crude carriers came into service making possible a reduction in the delivered costs of oil to Europe. As a consequence of these developments world shipyard tanker capacity has been fully booked for some time. It was only in the early months of 1971 that first indications of easing pressure on the tanker fleet were observed as demand for oil slackened owing to an economic slow-down in certain consuming areas. The tanker market eased so that preference could be given to long-haul instead of short-haul crudes. Some diversification in transportation has been sought through construction of pipe-lines to carry crude oil from the producing areas to ports on the Mediterranean. Although these pipe-lines offer an advantage in diversification, their location in potentially unstable areas implies that they are also a less secure form of transport.

163. When an area is dependent upon outside sources of supply for oil it is on economic grounds preferable to import this supply in the form of crude oil for refining in the area itself. Costs of transport are lower for crude oil compared to equivalent amounts of petroleum products. In some consuming countries refineries are demanded for reasons of regional or industrial policies, or for the profits available in this downstream activity. Moreover, there would appear to be more security in supply in imports of crude than in imports of products. A loss of a crude source can be made up more easily under today's market than can the loss of a product source. Finally, crude oil is considered less expensive to import (in terms of foreign exchange) than the equivalent in products, and for this reason, crude imports are more attractive for balance of payments reasons. OECD Europe has substantially moved toward self-sufficiency in refining capacity. Within the past two decades, the relative reliance of OECD Europe on imports of petroleum products has declined from 37 per cent of total supply in 1950 to about 5 per cent in 1970. Inasmuch as demand for products has roughly tripled during that time, this has meant that refining capacity has been established at an even greater rate.

164.   Despite the high costs of construction, additions to refinery capacity can be expected at least to keep pace with demand in the coming years. Local restrictions on refinery construction and the desire to hold industrial intrusion upon the environment to a minimum may complicate and delay expansion programmes, but none the less it is not regarded likely that product imports will increase to any great extent.

165.   Posted prices for crude oil remained basically unaltered throughout all the past decade, while the prices for the goods and services the host countries had to purchase from oil consuming countries were more or less caught up in an inflationary trend. The argument that purchasing power of the producers was being gradually eroded became a rallying point for OPEC action. Yet, total government receipts had increased sharply during the decade. Receipts for the Middle East and Libya, taken together rose from 75.6 cents per barrel in 1961 to 90.3 cents per barrel in 1970, and taking into account rising volumes of exports, total revenue for the Middle East producers alone in 1970 was almost triple the revenue nine years earlier.

166.   The shift from a buyer's to a seller's market in the year 1970 noted in paragraph 6 of this chapter, strengthened the negotiating position of the oil exporting countries. Pressures by the host governments for higher posted prices were first brought to bear in the autumn of 1970 as Libya, proceeding from its favoured geographic position relative to major consuming areas and outside the influence of the Suez Canal, sought and received higher posted prices and higher tax rates. Other producing countries followed suit. At the December 1970 meeting of OPEC a series of sweeping resolutions was adopted, in essence pointing to higher minimum tax rates and uniform general increases in tax reference prices. Shortly afterwards Venezuela unilaterally gave itself what the others were to achieve subsequently through negotiation. Algeria acquired in the beginning of 1971 a 51 per cent interest in all the oil operations inside its boundaries.

167.   Actions taken or contemplated by OPEC during the last months of 1970 led to negotiations between the host nations and the international oil companies which culminated in upward revisions of posted prices. These advances in posted prices were bartered against a promise of stability in pricing and security in supply through 1975. It was recognised that the pressures exerted on the oil companies by the host nations did mirror the realities of the market, that oil had been cheap in comparison with other forms of energy. And it did mirror the absence of any reasonable alternative which might have been grasped had some massive denial of oil taken place; i.e. the spare producing capacity once available in the Western Hemisphere, and which has been drawn upon in the past to make up for oil denied Europe, has disappeared. The ability to use the threat of a denial of oil as a tool in negotiations reflected an awareness by both parties that the market, which for so many years had favoured the buyer, had now changed dramatically in favour of the seller.

168.   The increase in posted prices will mean greatly expanded oil revenues for the producing nations. These new revenues will have to be provided by the consumer in the form of higher prices. Some of the new revenue may find its way back to the consumer nations in the form of additional orders for manufactured goods and commodities, thus lessening

74

any balance of payments difficulties. But oil should remain attractive compared with other forms of fuel and it is thought unlikely that its higher price will have a serious dampening effect on demand.

169.　　Oil-derived revenue for the Middle East and North Africa was about $5.9 billion in 1970, against a crude oil production of slightly more than 18 million barrels per day. By 1975, assuming that output might be in the order of 26.2 million barrels per day, which may well be conservative, government revenues from this oil will total almost $15 billion under those agreements initialled at Tripoli and Teheran.

170.　　The Geneva Agreement of January, 1972, which corrected the decline in the purchasing power of the US dollar, may well add an additional $1 billion to oil-derived revenue that year.

171.　　The specific details of the Libyan agreement on crude oil posted prices and taxes are illustrated in table 15; details of the Tehran agreement are presented in tables 16 and 17. Separate negotiations covered Iraqi crude oil moving via the IPC pipe-lines and Saudi Arabian crude delivered by the Trans-Arabian pipe-line to the Eastern Mediterranean. The results of these negotiations are given in tables 18 and 19.

172.　　Shortly after the Geneva agreement, OPEC took up the issue of participation with the companies operating in the Persian Gulf, thus dispelling any hopes that accommodation between industry and the host nations had been at last reached at least to 1975. The participation issue is discussed in paragraph 199, below.

B. THE INCREASING DEPENDENCE OF THE OECD AREA ON IMPORTED OIL SUPPLIES AND RELIABILITY OF THESE SUPPLIES

a) THE INCREASING DEPENDENCE

173.　　The three major oil-importing areas of the OECD— Western Europe, Japan and North America— have all been concerned with the question of security of supply and the possibility of diversification, either in sources of oil or in alternative uses of the several forms of primary energy. Diversification between sources on a geographic basis may also provide some political diversification, but there are no substantial quantities of oil supply available outside the OPEC group. The Soviet Union, although the second leading producer of crude oil in the world, has domestic and external obligations to meet with its own supplies, and once these obligations have been satisfied, only comparatively small amounts of oil are expected to be available for disposal in non-Communist oil markets.

174.　　Diversification between energy forms does not in fact offer much choice. The coal industry, being labour intensive, is continuously under the pressure of rising costs. In Europe this industry is declining. In the United States and some other areas in the world a revival of the coal industry has taken place because of favourable geological conditions and a massive infusion of labour-saving mining equipment. But in general the coal industry has not kept pace with the rise in energy demand. The advantages of natural gas, in cleanliness, ease of handling, and its expanding use as a raw material for the manufacture of petrochemicals, have generated a demand which cannot be covered by indigenous supplies; the combination of imports of gas by pipe-line and increasingly of liquefied natural gas has been, and remains, the only economically feasible alternative. But the importation of natural gas involves questions of cost, scale and, most important, flexibility of source.

175. Oil shale, tar sands, coal liquefaction and coal gasification must still be considered to be in the developmental stage. Similarly, nuclear power has yet to assume the role so confidently predicted for it; consequently, forecasts of the contribution of nuclear stations to power supplies in the coming years have become much more modest.

176. As far as energy is concerned, oil has been the mainstay of the dynamic economic growth in the OECD area, and there is little reason to doubt that most of the increase in energy supplies in the foreseeable future will also be covered by oil and by natural gas. While growth rates of the various industrialised economies are likely to slow down somewhat, reflecting the influence of a larger base upon which this growth must take place, the demand for oil will assume very large proportions. The magnitude of this growth can best be understood if expressed in absolute, rather than relative, terms. In sum the total annual demand of the OECD Member countries may increase by 1,215 million metric tons in the period 1970-1980. The allocation of this increment among North America, OECD Europe and Japan and the growth in imports required are summarized in table, page 90.

177. Clearly an increase in demand of this magnitude will mean, collectively for OECD an inevitable and expanded search abroad for supplies. Recent discoveries or established indigenous fields will by no means meet the requirement. Although the North Sea promises to be a major producing area it is clear from the figures above that its potential is not likely to allow a marked reduction in OECD Europes' imports. Moreover, the search for oil and gas on the continental shelf around the world may be impeded by unresolved questions relating to national jurisdiction, to environmental protection, and to the high costs of such undertakings. In the Western Hemisphere, the rich finds on the North Slope of Alaska remain unexploited as governments, producers, consumers and environmentalists seek to find an acceptable accommodation of the many diverse and seemingly incompatible issues at hand.

178. It is believed that the expansion in indigenous supplies of oil within the OECD area will be sufficient to cover little more than thirty per cent of the indicated growth in demand during the present decade. Thus, North America, OECD Europe, and Japan together must find by 1980 an additional 925 million tons per annum of oil from foreign sources, if forecast demand is to be fully satisfied.

179. The emerging pattern is one of increased dependence upon imports. The OECD area as a whole will become more vulnerable as time passes; the relative decline in OECD Europe's reliance on foreign procurements will be of little relevance. Japan has virtually reached the stage of full

76

	1970	1980
OECD dependence on imports	60.5	67.3
North America	21.6	39.5
OECD Europe	96.4	83.9
Japan	99.8	99.6

reliance on imports, and North America will have to look increasingly beyond its frontiers for oil needed to meet the growing gap between demand and indigenous supplies.

180.   Energy policies, and related political actions and programmes in the present decade will have to take into account the implications of this presumed enlarged dependence upon foreign supplies.   Previously, North America, because of its general self-sufficiency, was able to approach its imports of fuel from a different vantage point than could OECD Europe and Japan.   At that time North America was motivated not by need but more by economic and national security considerations.   Now, North America has lost this safety of self-sufficiency and the growth in imports during the 1970s may come very close to matching the growth in indigenous production.   Thus, North America can be expected to join OECD Europe and Japan in a common concern and a common evaluation of imports and imports policies as the 1970s unfold.

b) THE RELIABILITY OF SUPPLIES

181.   This growing dependence of the OECD area as a whole on oil imports points to the importance of the reliability of these imports over the next decade.   Since the publication of "Oil Today" in 1964, successive events have called attention to the vulnerability of oil supplies, especially for Western Europe, although none of these events actually led to serious shortages.   When an evaluation of future risks is attempted it is necessary to focus on possible causes of an interruption of supplies and the possibilities for mitigating or eliminating its effects.

182.   The basic cause of an interruption of supplies can be technical, commercial or political.   The first does not seem of major importance because the effects would be narrow and of short duration.   Commercial or political causes, however, can have a serious impact.   Political causes are difficult to predict but it may be said that in most oil exporting areas there is always a danger of political conflict.   The risk of commercial causes for an interruption in supplies seems to have increased.   In 1971 for the first time OPEC members formed a common front.   Circumstances in the oil market were such that an interruption of supplies by OPEC would have had a direct impact on the supply position of consuming countries.   A tighter oil market in the future is to be expected and this means more than ever that consuming countries will be vulnerable to supply interruption.   In short, the danger of interruptions of normal supplies will be greater in the years to come than in the past.

183.   Interruptions in supplies can result from interruptions in production or in transport routes (especially pipelines while the Suez Canal is closed) or in both.   In the case of an interruption in production, if no spare capacity is available, there are no means to mitigate the effects other than the use of whatever stocks are available.   In practice there will always be some spare producing capacity, but now that the surplus capacity of the United States has disappeared, the amount available outside the OPEC area is very small indeed.   In the event of an interruption in short-haul crude supplies, a transport crisis could well develop, as reserve tanker capacity may not be sufficient to handle the additional quantities of long-haul crude. For the same reason, and interruption involving the closure of Eastern Mediterranean pipelines will cause a tanker transport crisis.

184.   Normally there will be some reserve tanker capacity; but in most cases this reserve will not be great because supply and demand in the tanker market will tend to be in equilibrium.   The conclusion is that although there will almost always be some margin in flexibility in production and transport, this margin will structurally be insufficient to balance a major interruption in supplies.   Oil imports cannot be looked upon as being inherently reliable.   The implications to be derived from this conclusion will be examined in the next part of this chapter.

C. DEVELOPMENTS IN GOVERNMENT POLICIES ON OIL AND OIL SUPPLY

THE OIL PRODUCING COUNTRIES

185.   The 1960s saw a marked change in the international oil industry. New independent oil companies acquired an interest in overseas oil production, the Organisation of Petroleum Exporting Countries was formed, and national oil companies grew in importance in both consuming and producing countries.   With all countries and companies increasing production, the available supplies soon exceeded demand and inevitably prices declined. At the same time, producing countries began to seek revised terms for new agreements.   Some concession agreements contained terms giving host governments increasing control over the exploitation of their resources, and in many cases host governments required participation interests for their own oil companies in the new concessions.   Throughout the decade, however, changes in the basic arrangements occurred slowly.   In the following paragraphs we review the major changes during the 1960s in the legal relationship between the governments of oil producing countries and the oil companies.

186.   As governments have sought greater control over the natural resources of their countries, there has emerged a new form of legal relationship between the governments of oil producing countries and the oil companies. Instead of granting concessions to the oil companies with the right to explore for and to exploit petroleum on their own account, governments have vested these rights in their own national oil companies which have then entered into arrangements with the oil companies whereby the latter act as contractors, or operators, for the national oil company.   These arrangements are to be distinguished from conventional contractural agreements in

which the contractor is rewarded for his services by stipulated payments: his reward, and the scale of his reward, is entirely dependent upon the success of the venture; the contractor is required either to finance, or to introduce third parties to finance, the national oil company's share of development capital and operating costs; and the national oil company participates in the formulation of operating programmes and budgets.

187.   Although the details vary from contract to contract, four main types of national oil company— oil contractor company arrangements may be noted.   These types of arrangements are: a) joint venture arrangements; b) production sharing contracts; c) ERAP-type contracts; and d) Venezuelan-type service contracts.   Each is discussed in some detail in the Annex at the end of this chapter.

188.   At the beginning of the 1970s the pace of change in relationship between the international companies and the host nations accelerated markedly, for those reasons noted in paragraph 157 above.

189.   Libya for one was quick to recognize the advantage of its proximity to the major European market and reduced its production, departing from the practice of maximizing production as the means to maximize revenues which heretofore had characterised attitudes in the Eastern Hemisphere. Libya's action was followed by new demands from OPEC which eventually led to new agreements with all of the major oil producing countries.   Representing some 50 per cent of the world's oil production and accounting for 90 per cent of world crude oil exports, with little competition from other energy suppliers, OPEC countries have had the potential for great power for some time.   The combination of circumstances in 1970, however, highlighted the strong position of the producing countries, and OPEC found a new cohesiveness which enabled it to use its strength to advantage.

190.   OPEC had its genesis in the late 1950s as the major crude oil producers, all developing countries, with economies largely dependent on the export of this one product, became aware of their common interests in relation to those of the major international oil companies.   In September 1960, following the reduction of posted prices in the Middle East, representatives of Iran, Kuwait, Saudi Arabia, Iraq and Venezuela met in Bagdad and decided to join together in a permanent organisation to be called the Organisation of Petroleum Exporting Countries.

191.   The main objectives of OPEC, as described in its statues, are to: "co-ordinate and unify the petroleum policies of Member countries and determine the best means for safeguarding their interests, individually and collectively"; and to "devise ways and means of ensuring the stabilization of prices in international crude oil markets with a view to eliminating harmful and unnecessary fluctuations".

192.   In moving toward these objectives the OPEC countries developed a series of guidelines which were adopted as resolutions in subsequent conferences.   The most important of these resolutions dealt with the restoration of prices to their pre-August 1960 level, the "expensing of royalties"[1],

---

1.   The inclusion of royalties in the costs to be offset before calculating tap.

production quota schemes, and the application in all countries of a taxation system based on posted prices rather than actual prices. These objectives, with the exception of a programme for production quotas which has been opposed by several countries, have now been largely accomplished. Over the years membership in OPEC has been expanded to include, in addition to the founding five countries, the following six countries: Qatar, Libya, Indonesia, Abu Dhabi, Algeria, and Nigeria.

193.   Changes now being demanded by the producing countries have also been mooted for some time. At its Sixteenth Conference in June 1968, OPEC countries drew up a set of objectives to guide the long-term development of their petroleum resources. The following concepts were emphasized:
— the desirability of direct exploitation of resources by agencies of the producing countries, rather than by outside agencies;
— the doctrine of "changing circumstances" which is said to justify countries in demanding changes in the terms of existing concession agreements;
— under the principle of changing circumstances, the right to acquire a participating ownership share for governments under the existing concession agreements;
— the right of governments to alter the financial terms of agreements where companies are receiving "excessively high net earnings";
— the right of governments to determine posted or tax reference prices;
— the accelerated relinquishment of concession territories.

194.   OPEC conferences since that time have been directed towards means of achieving these objectives. The Twenty-First Conference, held in Caracas in December 1970, resulted in collective demands by the OPEC countries for alterations in the existing financial agreements, which led to the new agreements signed in Tehran and Tripoli in the first half of 1971.

195.   Estimates place the increased revenues won by the oil exporting countries, as a result of these agreements, for the period 1971-75 at more than $30 billion. A subsequent OPEC meeting, held in Vienna in July 1971, concentrated on measures to obtain increased participation for the producing countries in the operations of the oil companies.

196.   The OPEC meeting in Beirut on 22nd September, 1971, resulted in the adoption of resolutions calling specifically for negotiations between the Member countries and the oil companies on two subjects:
1.   compensation to the producing countries for any loss of income caused by changes in the world monetary situation, and
2.   acquisition by the producing countries of a participating share in the production activities of the concessionary companies. The resolutions are general in content, and the Member countries have been left free to follow whatever negotiating tactics or goals they may feel suitable.

197.   An agreement on the monetary issue was arrived at in Geneva in January 1972 for Middle East oil and for oil delivered to the Eastern Mediterranean. By this agreement posted prices were increased by 8.49 % and provisions were made for future adjustments in parity rates.

198.   The desire for control over the exploitation of natural resources can be fulfilled in one of several ways. The most far reaching form is that of entire or partial nationalisation. An example of entire nationalisation is that of the Iraq Petroleum Company, whereas Algeria partly (51 per cent) nationalised the existing concessions. It may also take the form of government regulations, such as new legislation in Venezuela which imposes new restraints on company operations. A third form is that of participation, which may take the form of equity participation in existing concessions. The producing countries claim that the demand for partial ownership of the operating concessions is justified under the principle of "changing circumstances".

199.   In February 1972 discussions were initiated between the Persian Gulf members of OPEC and representatives of the international oil companies operating in those countries on the question of host nation participation in the crude oil producing operations of the companies. On 10th March the Arabian American Oil Company (Aramco) accepted the principle of 20 per cent Government participation in the company, with the questions of timing and method of acquisition of the 20 per cent interest, the price to be paid and the mechanism for payment, disposition of production, supply of future capital requirements and assurances for the future to be negotiated at future meetings.

A draft participation agreement between the international oil companies and the negotiator for the Persian Gulf states, Sheikh Ahmad Zaki Yamani, the Oil Minister for Saudi Arabia, was initialled in New York City on 5th October, 1972. This draft agreement contained these guidelines: government ownership of the established producing concessions will start with 25 per cent in 1973, remaining constant through 1978, rising to 30 per cent in 1979 and 5 per cent more each year until 1982, and then by 6 per cent to 51 per cent in 1983. Compensation paid to the companies will be based on a so-called "updated book value". The amounts of participation oil accruing to the governments will be divided into three categories. A portion will be set aside for direct sale by the state companies, a portion will be sold back to the companies under long-term contracts and a portion will be used to balance out overliftings and underliftings. The precise buy-back prices will be a matter of negotiations between the individual countries and their concessionaries.

At the XXX (Extraordinary) meeting of OPEC on 26th-27th October, 1972, the four Persian Gulf states— Saudi Arabia, Kuwait, Abu Dhabi and Qatar —declared their intent to become parties to this draft agreement. The fifth Gulf state, Iraq, reserved its position. It should be noted that the general agreement, to be valid, need be signed and ratified only be a minimum of three states.

200.   Not all OPEC members are interested in participation. Algeria and Venezuela have secured their interests in other ways, as is evident from the above. Indonesia has developed its own system of licences and production sharing, whereas Iran and the Consortium of oil companies in that country are working out an agreement of close cooperation during a long period. Not all OPEC members who are interested in participation have the same ideas about the amount of the initial percentage of participation. Some

may seek a higher percentage than the 20 per cent, accepted in principle by the main international oil companies.

201.   Whereas in the present system producing countries act mainly as collectors of ever rising royalties and taxes, the move in the direction of control and ownership will involve those countries more and more in the problems of the oil business.   Only to mention a few of these problems, there is the care of providing the necessary investment funds and the marketing of a highly competitive product.

THE OIL CONSUMING COUNTRIES

202.   Those principles which guided the OECD countries during the early part of the 1960s prevailed at the close of the decade and are likely to prevail in the years ahead.   They are the availability of energy in the forms and amounts desired, from secure sources and at the lowest reasonable cost to the consumer.   These are very broad principles indeed, and there are no consuming countries in the world that could not, nor would not, with some difference in accentuation, subscribe to them.   But the ease in statement of these principles is in no way matched by ease in fulfillment. Attempts to balance security of supply against rising costs are not always successful; neither are attempts to sacrifice security of supply in an effort to provide the consumer with cheaper energy.

203.   Oil has been a low-cost fuel material and it has served this purpose well.   But low-cost oil has meant foreign oil, at least for OECD Europe and Japan; indigenous supplies in these areas, meagre as they are, fall short of making meaningful contributions to total supplies.   Moreover, this indigenous oil is generally costly to produce and governments must weigh the advantages of low-cost oil against the spectre of political disturbances which may threaten or actually disrupt normal supply lines.   In the United States, during the sixties, the accent was put to a large degree on security of supply and a policy was followed of restricting imports through the Mandatory Oil Import Programme, with the attendant effect of maintaining in comparison with world prices, high crude prices at the well-head. Adherence to the oil import policy made it possible to keep the dependence of the United States on imported oil at a relatively low level.

204.   The realisation that oil will continue to be an essential energy source for some time yet, and that OECD's reliance on imported oil is increasing, combined with the apparent greater opportunity for collective action by producing countries all against the background of the supply interruption in 1967 and the fears of interruption in 1970 and 1971, have led most consuming countries to examine new means of assuring their requirements.   Moreover, the security which had been provided for many years by excess capacity outside the Eastern Hemisphere had disappeared by the end of the 1960s.   The United States today imports oil in amounts which could not be made up from domestic sources in the event of foreign supply interruptions.   Reserves in Venezuela have declined steadily since 1965 and Venezuela, as the decade ended, was selling all the oil it could produce.   Now the continuing under-current of unrest in the Middle East and North Africa and the ever-present spectre of armed conflict which

82

threatens the orderly flow of oil from the producer to the consumer sharpen the need to focus on the security aspects of oil supply.

205.　　Oil is merely one part of the energy spectrum; other fuels, especially gas, nuclear energy and in some cases coal warrant attention as well.　This diversification among the various fuel materials plays an important role in the security problem.　Governments must seek to create conditions which maintain or lead to investment in alternative fuels.　Because a majority of these fuels— nuclear energy, shale oil, tar sands, liquefaction or gasification of coal— require tremendous lead times and vast infusions of investment capital, this policy has to be forward looking.

206.　　Discussions in the OECD Oil Committee of means of insuring against supply interruptions over the short term have centered on stockpiling programmes.　Other factors, including diversification, were also considered, but such factors have their effect only in the long term.　In the future more attention will be devoted to an elaboration of pre-crisis preparedness.

207.　　In July of 1962 the Council of the OECD recommended an oil stockpiling programme to its European Member countries.　It had as its basic recommendation a minimum stock level equal to 60 days' current inland consumption or 65 days of the inland consumption in the previous year.　It further noted that Member countries could not, in the event of an emergency, call upon the stocks of another country accumulated prior to the emergency and that the cost of each national stockpiling programme had to be borne by the country concerned.　This programme sufficed for the greater portion of the past decade.　But the cumulative effect of a number of developments' in the oil supply situation, noted in previous paragraphs, led the Oil Committee to conclude that the recommended 65-day minimum was no longer adequate.　In July of 1971 the Council recommended to the governments of the European Member countries that steps be taken to achieve as soon as possible a stock level of at least 90-days average inland consumption of the previous calendar year for motor gasolines and aviation fuels, gas/diesel-oil, kerosene and kerosene type jet fuels and fuel-oils.

208.　　Further consideration will be given in the future to the stockpiling programmes.　At the moment the 90-day recommendation is directed only to the European Member countries (except Finland).　Japan, however, is virtually in the same supply position.　The stockpiling programme of Japan currently in force has set a target of 65 days to be achieved by 1st April, 1975.　Also, the United States has become in absolute terms a major oil importing country.　Thus, security stocks are increasing in importance for the OECD area as a whole.　The significance of stocks lies not only in their value after a crisis has broken out.　Their known presence may also have a preventive effect and lessen the risks of a crisis.

209.　　The Tehran and Tripoli negotiations involved, for the first time, collective negotiations on both sides.　While future negotiations between producing countries and companies will probably continue to be collective in nature, the question of further involvement on the part of consumer governments has also been raised.　Moreover, the advisability of closer relations between consumer and producer governments may also be explored. Yet, while the future role of consumer governments may not be clear at this juncture, it is certain there will be a demand for more effective consultation procedures than have been available to date on energy issues affecting consumer national interests. Also, the interests of the consuming countries will be more closely harmonised. All these actions have to be worked out against the background of an absolute need for close cooperation between the consuming countries in the light of the close cooperation that has been developed by the producing countries.

210.   It has become evident that a policy on oil as an energy source in itself will not be sufficient; viable policies embracing every form of primary energy must be developed, in harmony with one another. Recognizing this, consuming countries have turned to studies of programmes which would encourage the development of domestic hydrocarbon resources, accelerate the development of nuclear energy, and reconsider the role of coal in total energy requirements. While such measures may increase the costs of energy, they may strike a balance which is more consonant with the general national interest of the consuming countries.

Table 15. LIBYAN AGREEMENT ON POSTED PRICES AND TAXES, MEDITERRANEAN, 1970-1975
(Dollars per Bbl.)

	PRE-SEPTEMBER 1970	1st JANUARY 1971	20th MARCH 1971	1st OCTOBER 1971	1st JANUARY 1972	1st JANUARY 1973	1st JANUARY 1974	1st JANUARY 1975
Base Posted Price (40° gravity)	2.23	2.550	3.197	3.197	3.217[1]	3.368	3.523	3.682
Temporary – Suez Premium	-	-	0.120	0.120	0.120	(*)	(*)	(*)
Temporary – Freight Premium[2]	-	-	0.130	0.082	0.082[3]	(*)	(*)	(*)
Total Posted Price	2.23	2.550	3.447	3.399	3.419	3.368	3.523	3.682
Royalty 12.5% of Posted Price	0.28	0.318	0.431	0.425	0.427	0.421	0.440	0.460
Average Producing Cost[4]	0.30	0.300	0.300	0.300	0.300	0.300	0.300	0.300
Tax Reference Price	1.65	1.927	2.716	2.674	2.692	2.647	2.783	2.922
Tax[5]	0.83	1.060	1.494	1.471	1.481	1.456	1.531	1.607
Government Take:								
Tax	0.83	1.060	1.494	1.471	1.481	1.456	1.531	1.607
Retroactive Buy-Out[5]	-	-	0.090	0.090	0.090	0.090	0.090	0.090
Royalty	0.28	0.318	0.431	0.425	0.427	0.421	0.440	0.460
Total Government Take	1.11	1.378	2.015	1.986	1.998	1.967	2.061	2.157
Company's Tax-Paid Cost:								
Total Government Take	1.11	1.378	2.015	1.986	1.998	1.967	2.061	2.157
Average Producing Cost	0.30	0.300	0.300	0.300	0.300	0.300	0.300	0.300
Total Company Cost	1.41	1.678	2.315	2.286	2.298	2.267	2.361	2.457
Rise in Total Postings	-	0.320	1.217	1.169	1.189	1.138	1.293	1.452
Rise in Government Take per Bbl.	-	0.268	0.905	0.876	0.888	0.857	0.951	1.047

* Assuming temporary premiums remain in full effect 1971 and 1972 but not thereafter.

1. Includes 2 cents sulfur-escalation under September 1970 deal.
2. Temporary freight premium based on following world scale rates: First quarter index W 94,4; 2nd quarter, W 90,3; 3rd quarter, W 86,1.
3. Assumes temporary freight advantage in January 1972 same as October 1971,
4. Producing costs reported at 25 cents to 26 cents by the large-volume producers, ranging up to 50 cents for smaller volumes, A per barrel industry average of 30 cents assumed here,
5. Tax Rates: 50% before September 1970, Since september variable tax rates of 54%-58% applicable to different companies, averaging about 55%, in lieu of lump sum payment of "retroactivity" settlement, New base tax rates rises from former 50% to 55%, and each company will "buy out" its last September tax "retroactivity" settlement with an additional per barrel payment, averaging about 9 cents, See note on Tripoli Agreement,

Table 16. IRANIAN AGREEMENT ON POSTED PRICES AND TAXES, 1970-1975,- PERSIAN GULF SHIPMENT

(Dollars per Barrel)

	PRE-NOV. 14 1970	AFTER-NOV. 14 1970	FEB. 15 1971	JUNE 1 1971	JAN. 1 1973	JAN. 1 1974	JAN. 1 1975
Base Posted Price (34° gravity)	1.790	1.790	2.170	2.274	2.381	2.491	2.603
OPEC Allowances and gravity adjustments	0.099	0.099	-	-	-	-	-
Royalty 12.5%	0.224	0.224	0.271	0.284	0.298	0.311	0.325
Average Producing Cost	0.120	0.120	0.120	0.120	0.120	0.120	0.120
Tax Reference Price	1.347	1.347	1.779	1.870	1.963	2.060	2.158
Tax[1]	0.674	0.741	0.978	1.029	1.080	1.133	1.187
Government Take:							
Tax	0.674	0.741	0.978	1.029	1.080	1.133	1.187
Royalty	0.224	0.224	0.271	0.284	0.298	0.311	0.325
Total	0.898	0.965	1.249	1.313	1.378	1.444	1.512
Company's Tax Paid Cost	1.018	1.085	1.369	1.433	1.498	1.564	1.632
Rise in Posted Price	-	-	0.380	0.484	0.591	0.701	0.813
Rise in Government Take	-	0.067	0.351	0.415	0.480	0.546	0.614

1. Tax rate at 50% until Nov, 14, 1970, then increased to 55%.

NOTE: The Teheran Price Settlement: A 5-year price agreement was reached by the producing companies and the six Persian Gulf States of Abu Dhabi, Iran, Iraq, Kuwait, Qatar, and Saudi Arabia and established a new basis for pricing and taxing for the period February 15, 1971 to December 31, 1975. Essentially the agreement was as follows: (a) a tax rate of 55%, (b) abolition of all prior OPEC allowances and adjustments effective February 15, 1971; (c) a basic 33 cents per bbl, increase in posted prices; (d) a further 2 cents per bbl, in consideration of "freight disparities"; (e) a 5 cents per bbl, per degree API for each degree below 40, 0° to 40, 9° API to be added with an additional 1 cent per bbl, to oil in the 30, 0° to 31, 9° API range. Done in 1° API increments; (f) an exceptional 6 cents per bbl, for Iraq Basrah crude to account for royalty not being expensed; (g) automatic increases on June 1, 1971, January 1, 1973, 1974 and 1975 by 2.5% of the prevailing posted price prior to date of increase plus a flat 5 cents per bbl, added. The 2.5% increase to be calculated to nearest 1 cent. These increases are intended to protect against inflation; (h) some additional protective covenants against subsequent settlements in the Mediterranean which would create a freight disparity between Persian Gulf and Mediterranean prices,

86

Table 17. SAUDI ARABIAN AGREEMENT ON POSTED PRICES AND TAXES, 1970-1975 – PERSIAN GULF SHIPMENTS
(Dollars per Barrel)

	PRE-NOV. 14 1970	AFTER NOV. 14 1970	FEB. 15, 1971	JUNE 1, 1971	JAN. 1, 1973	JAN. 1, 1974	JAN. 1, 1975
Base Posted Price (34° gravity)	1.800	1.800	2.180	2.285	2.392	2.501	2.614
OPEC Allowances and gravity adjustments	0.099	0.099	-	-	-	-	-
Royalty 12.5%	0.255	0.255	0.273	0.286	0.299	0.313	0.327
Average Producing Cost	0.120	0.120	0.120	0.120	0.120	0.120	0.120
Tax Reference Price	1.326	1.326	1.787	1.879	1.973	2.068	2.167
Tax[1]	0.663	0.729	0.983	1.033	1.085	1.137	1.192
Government Take:							
Tax	0.663	0.729	0.983	1.033	1.085	1.137	1.192
Royalty	0.255	0.255	0.273	0.286	0.299	0.313	0.327
Total	0.918	0.984	1.256	1.319	1.384	1.450	1.519
Company's Tax Paid Cost	1.038	1.104	1.376	1.439	1.508	1.570	1.639
Rise in Posted Price	-	-	0.380	0.485	0.592	0.701	0.814
Rise in Government Take	-	0.066	0.338	0.401	0.466	0.532	0.601

1. Tax rate at 50% until Nov. 14, 1970, then increased to 55%.

NOTE: The Teheran Price Settlement: A 5-year price agreement was reached by the producing companies and the six Persian Gulf States of Abu Dhabi, Iran, Iraq, Kuwait, Qatar and Saudi Arabia and established a new basis for pricing and taxing for the period February 15, 1971 to December 31, 1975. Essentially the agreement was as follows: (a) a tax rate of 55%; (b) abolition of all prior OPEC allowances and adjustments effective Feb. 15, 1971; (c) a basic 33 cents per bbl, increase in posted prices; (d) a further 2 cents per bbl, in consideration of "freight disparities"; (e) a 5 cents per bbl, per degree API for each degree API below 40, 0° to 40.9° API to be added with an additional 1 cent per bbl, added to oil in the 30, 0° to 31, 9° API range. Done in 1° API increments; (f) an exceptional 6 cents per bbl, for Iraq Basrah Crude to account for royalty not being expensed; (g) automatic increases on June 1, 1971, January 1, 1973, 1974 and 1975 by 2.5% of the prevailing posted price prior to date of increase plus a flat 5 cents per bbl, added. The 2.5% increase to be calculated to nearest 1 cent. These increases are intended to protect against inflation; (h) some additional protective covenants against subsequent settlements in the Mediterranean which would create a freight disparity between Persian Gulf and Mediterranean prices,

Table 18. IRAQ AGREEMENT ON POSTED PRICES AND TAXES AT EASTERN MEDITERRANEAN (VIA IPC-LINE) 1970-1975 (ESTIMATED)

(Dollars per Barrel)

	PRE-SEPTEMBER 1970	1st JANUARY 1971	20th MARCH 1971	1st OCTOBER 1971	1st JANUARY 1972	1st JANUARY 1973	1st JANUARY 1974	1st JANUARY 1975
Base Posted Price (36° gravity)	2.210	2.410	2.971	2.971	2.971	3.095	3.222	3.353
Temporary - Suez Premium	-	-	0.120	0.120	0.120	(*)	(*)	(*)
Temporary - Freight Premium[1]	-	-	0.120	0.075	0.075[2]	(*)	(*)	(*)
Total Posted Price	2.210	2.410	3.211	3.166	3.166	3.095	3.222	3.353
Royalty 12.5%[3]	0.276	0.301	0.401	0.396	0.396	0.387	0.403	0.419
Average Producing Cost[4]	0.120	0.120	0.100	0.100	0.100	0.100	0.100	0.100
Tax Reference Price	1.814	1.989	2.711	2.670	2.670	2.608	2.719	2.834
Tax[5]	0.907	0.994	1.491	1.469	1.469	1.434	1.495	1.559
Government Take (Est.)	1.183	1.295	1.892	1.865	1.865	1.821	1.898	1.978
Company's Tax Paid Cost	1.303	1.415	1.992	1.965	1.965	1.921	1.998	2.078
Rise in Total Postings	-	0.200	1.001	0.956	0.956	0.885	1.012	1.143
Rise in Government Take per Bbl.	-	0.112	0.709	0.682	0.682	0.638	0.715	0.795

*   Minimum - Assumes temporary elements fully eliminated.

1.   Temporary freight premium based on following world scale rates: First quarter index W 94, 4;   2nd quarter, W 90, 3; 3rd quarter, W 86, 1.

2.   Assume temporary freight premium same as for October 1971.

3.   Royalty is based on 12.5% of Posted Price - Negotiation for other considerations. For practical purposes same method of calculations used as for other Mid-East producing countries since Iraq has a different method for calculating taxes and uses different bases.

4.   Approximately 12 cents bbl, but changes, in expensing and capitalization accounts, may reduce it to 10 cents per bbl, after 1970.

5.   Formerly 50% of profits. Profits were calculated at border price less producing costs, Royalty not expensed, For practical purposes will estimate using 55% rate and expensing royalty a la OPEC. May be an acceptable estimation,

See note on Tripoli Agreement,

88

Table 19.   SAUDI ARABIE AGREEMENT ON POSTED PRICES AND TAXES AT EASTERN MEDITERRANEAN (VIA TAPLINE), 1970-1975)

(Dollars per Barrel)

	PRE-SEPTEMBER 1970	1st JANUARY 1971	20th MARCH 1971	1st OCTOBER 1971	1st JANUARY 1972	1st JANUARY 1973	1st JANUARY 1974	1st JANUARY 1975
Base Posted Price (34° gravity)	2.170	2.370	2.941	2.941	2.941	3.065	3.192	3.322
Temporary - Suez Premium	-	-	0.120	0.120	0.120	(*)	(*)	(*)
Temporary - Freight Premium[1]	-	-	0.120	0.075	0.075[a]	(*)	(*)	(*)
Total Posted Price	2.170	2.370	3.181	3.136	3.136	3.065	3.192	3.322
Royalty 12.5%	0.271	0.296	0.398	0.392	0.392	0.383	0.399	0.415
Average Producing Cost	0.120	0.120	0.120	0.120	0.120	0.120	0.120	0.120
Tax Reference Price	1.779	1.954	2.663	2.624	2.624	2.562	2.673	2.787
Tax[3]	0.890	0.977	1.465	1.443	1.443	1.409	1.470	1.533
Total Government Take	1.161	1.273	1.863	1.835	1.835	1.792	1.869	1.948
Company's Tax Paid Cost	1.281	1.393	1.983	1.955	1.955	1.911	1.989	2.068
Rise in Total Postings	-	0.200	1.011	0.966	0.966	0.895	1.022	1.152
Rise in Government Take per Bbl.	-	0.112	0.590	0.674	0.674	0.631	0.708	0.787

*   Minimum - Assumes temporary elements fully eliminated.

1.   Temporary freight premium based on following world scale rates: First quarter indew W 94.4; 2nd quarter, W 90.3; 3rd quarter, W 86.1.

2.   Assumes temporary freight advantage in January 1972 same as October 1971.

3.   Tax rate at 50% prior to March 20, 1971, and at 55% thereafter.

NOTE: The Tripoli Agreement.   This agreement affects price and tax rate changes of Libyan oil and other OPEC countries that ship oil from Mediterranean ports.   For Libya only a "base" posting.   effective March 20, 1971, sets posted price at $3.07 for 40° API oil with 2 cents added for each.   1° API above 40° and decrease by 15 cents per each.   1° API below 40.09° API.   This base posting includes a 10 cents low sulphur premium which increases 2 cents per bbl.   per annum on 1/1/72, '73, '74 and 1975.   The annual inflation factor escalation is the same as for the Teheran settlement of 2.5% + 5 cents per bbl, at sepcific dates.   Additional freight premiums are as follows: Suez premium 12 cents per bbl, added.   If canal opened to 37' depth, 8 cents of this will be eliminated.   if 38' depth reached remaining 4 cents per bbl, would be eliminated.   Temporary freight premiums calculated on basis of average quarterly tanker freights.   Rate until June 1, 1971 is set at 13 cents per bbl.   Taxes increase to 55% with certain "retroactivity" payments made to adjust.   Will average out at around 9 cents per bbl, for most companies.   The temporary freight premiums will be calculated at 0.058 cents per bbl, in Libya for each 1 percentage point of WS by which LR2 AFRA exceeds WS 72 but no discount allowed below WS 72.   The rate for Saudi Arabia and Iraq are 0.053 cents per bbl., a locational variance to Libya's 0.058 cents/bbl.

89

	1970	1980	INCREMENT
Demand			
North America	764	1,209	445
OECD Europe	619	1,113	494
Japan	190	466	276
Total	1,573	2,788	1,215
Indigenous Supply			
North America	599.4	731	131.6
OECD Europe	22	179	157
Japan	0.8	2.3	1.5
Total	622.2	912.3	290.1
Imports Required			
North America	164.9	478	313.1
OECD Europe	597	934	337
Japan	189.6	464	274.4
Total	951.5	1,876	924.5

EXAMPLES OF NATIONAL OIL COMPANY-CONTRACTOR ARRANGEMENTS

For convenience in discussing these arrangements, the "oil company" will hereafter be referred to as the "contractor".

a) JOINT VENTURE ARRANGEMENTS

This title may suitably describe a relationship in which the national oil company and the contractor have status more nearly equal, albeit with different participating interests, in the joint venture than in the other forms described below. A typical recent example is the contract between the Libyan National Oil Company (NOC) and Shell Exploration (Libya) Ltd. (SELL). The participating interests of NOC and SELL are scaled according to levels of production beginning with 25 per cent NOC and 75 per cent SELL until production reaches a level of 260,000 b/d, to 50 per cent SELL when production reaches 500,000 b/d.

All exploration expenditures are borne solely by SELL, which is also required to advance NOC's share of development capital and operating expenditure against reimbursement out of 50 per cent of NOC's share of crude oil produced.

SELL will pay in accordance with the Petroleum Law royalty for its share of production and taxes on its profits from the joint venture.

b) PRODUCTION SHARING CONTRACTS

These have now become the standard form of arrangement between the Indonesian State Oil Company Pertamina and its contractors. All the operations are financed by the contractor which may retain up to 40 per cent of the annual production to cover amortization of capital costs and operating expenses. The balance of production is then divided between Pertamina and the contractor, in some cases in a fixed proportion of 65/35, in others rising to 70/30 above certain levels of production. Pertamina pays all royalties and taxes to the government, including taxes on the profits of the contractor arising from the production-sharing contract.

A similar form of arrangement has been adopted by Petroperu in Peru, except that the production is divided equally between Petroperu and contractor, the contractor recovering all capital and operating costs and his remuneration out of its share of production, and Petroperu paying out of its share of production all royalties and taxes, including the taxes on contractor's profits from the operations.

c) ERAP-TYPE CONTRACTS

In December of 1966, the French State Oil Company ERAP (Entreprise de Recherches et d'Activités Pétrolières) entered into a novel-type arrangement with the National Iranian Oil Company (NIOC). All the operations are to be financed by ERAP until NIOC is in a position to finance them out of the cash flow derived from operations under the agreement. The advances by ERAP are to be treated as loans to be repaid only if commercial discovery is made and then only from the time commercial production commences.

Fifty per cent of discovered reserves is to be set aside for subsequent development by NIOC: the remaining 50 per cent is to be developed under the terms of the NIOC/ERAP agreement. ERAP's remuneration consists of the right for a period of 25 years to purchase fob the export terminal up to 35-45 per cent (depending on the length of the pipe-line from field to export terminal) of the production developed under the Agreement at the tax-paid cost (the tax element being based on realized prices). ERAP is exempted from taxation. ERAP, as broker for NIOC, undertakes to export to world markets part of the remainder of the production from fields developed under the agreement at realized prices less a 2 per cent commission; the proceeds of these sales are to be utilized by Iran for the purchase of French equipment, products or services. In 1968, the Iraq National Oil Company entered into a similar contract with ERAP, which differs from the above only in details.

d) VENEZUELAN-TYPE SERVICE CONTRACTS

The three contracts recently concluded by the Corporacion Venezolana de Petroleo (CVP) for the development of blocks having high prospective value in South Lake Maracaibo are of a more sophisticated nature containing, as they do, provision for the making by the contractor of additional payments to CVP related to its net profits. The following are the highlights of the CVP/Shell Surca contract.

All the operations are to be financed by the contractor. After a three-year exploration period the contractor may retain only 20 per cent of the acreage, consisting of the first and third choice of 50 square kilometre blocks each, CVP having the second choice (and thereby probably including part of any structure containing a commercial discovery) and retaining the remaining 70 per cent.

The contractor retains 90 per cent of crude oil production, CVP the balance.

The contractor pays a government royalty of 16-2/3 per cent and an income tax currently 60 per cent computed on the basis of Fiscal Export Prices for crude and products. In addition, a payment will be made to CVP of a royalty equal to 5 per cent of the conventional royalty. A payment will also be made to CVP of a proportion of contractor's after-tax profit on a sliding scale, from zero to 10.4 US cents/bbl for the net profit range of 18-50 US cents/bbl. Net profits excess to 50 US cents/bbl will engender payments of 55 per cent of the excess.

CVP has participation in management through joint operating committees and has an option to take up to 20 per cent of shareholding in the contractor company, CVP paying its share of past and future costs; this option is exercisable upon the selection of exploitation acreage.

V

THE REFINING INDUSTRY

A. GENERAL INTRODUCTION

211.   Before World War II the demand for oil products was still modest and the range of products not very extensive. Motor gasoline was the dominant product in the market, and quality specifications were far less complex than they have become since. Under these circumstances advantage of the economies of scale could best be taken by concentrating refining in the crude exporting countries. After the war a number of factors brought about a change in this pattern. A trend towards the construction of refineries close to the area of consumption rather than close to the crude source, already apparent before 1960, continued to accelerate during the 1960s. The main factors behind this trend were:

— the rapid increase in demand for a variety of refined products, making the construction of such refineries economic;

— the development of large tankers, favouring shipment of crude oil rather than the more costly shipment of products in smaller tankers;

— lessening the foreign exchange costs, yet satisfying expanding oil needs, by importing the less expensive crude instead of oil products;

— from a strategic point of view the import of crude is somewhat more secure than the import of refined products;

— the economic advantages implicit in the refining industry of processing a package of crudes of different types with complementary qualities close to market, compared to refining at crude oil producing centres where generally less intake flexibility is available.

The construction of refineries in consuming areas has also been favoured by governments by the imposition of import duties on finished products.

212.   This trend does not suggest that refineries in producing countries have no place in the current oil supply market. Such refineries in fact play a very useful "balancing" role in supplying the import requirement for feedstocks, special products, blending components and finished products, to a large variety of markets, other than their own. The growth of refining capacity in two main oil producing areas— the Caribbean and the Middle East— from 1960 to 1970 is shown in Table 20. In these ten years, refining capacity in the Caribbean enlarged by 120 per cent, while that in the Middle East expanded by only 87 per cent. These growth rates are rather modest in relation to those for Japan and Western Europe, which were 450 per cent and 200 per cent respectively. Apart from the "traditional" producing

areas, relatively small refineries have been built in "new" producing countries such as Algeria, Libya and Nigeria, which refineries at this stage are geared to local demand only.

213.   Table 21 depicts the growth in refinery capacity in the OECD area during the last decade, and projects capacity forward to 1975.   As may be seen the relative growth in Japan has been very striking.   The growth in Europe however, has been more substantial in terms of tons added during the decade.   For every ton in capacity built in Japan, almost four tons were built in Europe.   Apart from a larger growth in oil consumption, the reason for the difference in growth rate between Japan and Europe is that the build-up of refining capacity in Japan proceeded against a much smaller base than that in Europe.   The growth in refinery capacity in North America during the last decade has been comparatively modest, namely 39 per cent. Refinery development there had undergone its most rapid expansion in the years prior to 1960.   It should be stressed that in the more recent years the United States, for a variety of reasons set out under paragraph 215, has met much of its growth in domestic demand for fuel-oils by the construction of refining capacity outside the country.

214.   With respect to the pattern of refining, two factors are worthy of mention: the economy of scale and the meeting of product demand by a combination of crude oil selection and processing equipment.   The oil industry has a strong economic incentive to build primary and secondary processing units as large as can be justified by the circumstances at each location.   As a consequence there is at least in Europe an evident tendency for the smaller— and more particularly, inland— refineries to produce a balanced output to meet local demand, and for the large coastal refineries to have more complex and specialized plant, capable of meeting deficit demand and absorbing surplus products from their "satellites".

215.   In discussing the patterns of refining, it is perhaps more appropriate to consider the several OECD areas separately.   In North America, and especially in the United States (as implied by the data in Table 22) the demand pattern leads to complex "deep" refining; high capital and operating expenditure are justified to upgrade low-valued residual material to gasoline and middle distillates.   In general, the US pattern emphasizes light ends, especially gasoline, in response to market demand, and de-emphasizes residual yields, with demand for this fuel met by imports.   Moreover, the economics of the refining operations in the United States are influenced by the import restrictions on light products and by the relative freedom to import fuel-oil.   Hence, the dependence of the US East Coast on fuel-oil imports from the Caribbean area where integrated oil/chemical complexes have been built to take advantage of this relative freedom.

216.   In Japan, the effect of government control of refinery expansion has been for the country to balance its supply and demand by importing petroleum products— in particular straight run naphtha and fuels.   The pattern of refinery output as shown in Table 22, based on crude oils imported predominantly from the Middle East, has not required refiners to build expensive conversion capacity to "balance the barrel".   To the extent that the rapidly growing petrochemical industry could not obtain feed-stock supplies from adjacent refiners, their needs have been met by imports.

94

217.   In Europe— as stated in Chapter III— the supply sources of crude oil to meet the rapid demand growth during the last decade showed a significant swing towards "short-haul" crudes from Libya, Algeria and Nigeria.   These crude oils are typically lighter than those from the Middle East (i.e. they contain more light and middle distillates and less residue).   It was therefore possible for the refining industry to meet the growing demand from the petrochemical industry for light distillate feedstocks relatively inexpensively, without the construction of costly conversion equipment. There was a tendency in Europe, except in Italy, for chemical manufacturers to locate their plants in the neighbourhood of large oil refining centres from whom feedstock supplies could be assured.

B. REFINERY DEVELOPMENT IN EUROPE

THE EXPANSION OF DISTILLATION CAPACITY

218.   European refinery capacity by the end of 1960 was 230 million tons and increased to 699 million tons by end 1970, a relative growth of 200 per cent during the period to a level at which total capacity exceeded total demand.   A further growth of 43 per cent, to 1,000 million tons by 1975 is estimated by Member countries.   On the basis of energy demand forecasts a further increase to about 1,300 million tons by 1980 may be expected.

219.   The largest absolute growth in refinery capacity from 1960 to 1970 took place in Italy, followed, in order, by Germany, France, the United Kingdom and the Netherlands; these five countries together account for 80 per cent of the total European capacity.   In percentage terms Spain and the Scandinavian countries showed a very rapid growth.

LOCATION OF REFINERY CAPACITY

220.   The expansion of trunk pipe-line capacity for crude oil led to a very rapid expansion in refining capacity inland in the first half of the past decade, as is illustrated in Table 23.   In the second half of that decade the ratio between coastal and inland refineries remained stable at around 70:30. There are of course major differences between the individual countries with size, geography and length of coastlines being decisive features.   (In Chapter VI, B, reference is made to the expansion of product pipe-lines in Europe.) While these features have stimulated the expansion of coastal refineries in large refining centres such as the Rotterdam and Marseille areas, as long as an inland market has a significant demand for heavy fuel there will be an incentive to build inland refineries supplied with crude oil by means of trunk pipe-lines, because the cost of transport per unit of heavy fuels over long distances is much greater than for crude oil.

221.   The trend to expansion at existing refinery locations has been very strong, and is likely to continue until the available land area places a limit on growth. For the years 1966-1970 the expansion at existing locations amounted to 156 million tons and the building of refineries at new sites added 94 million tons.   On the basis of the existing plans it is anticipated

that this trend will be further magnified during 1971-1975, as expansion at existing locations is to total 298 million tons, whereas construction at new sites is to add only 44 million tons. Table 24 illustrates how the average size of refineries has increased between 1965 and 1970. The largest refinery at present has a capacity of 25 million tons per year. This growth in average refinery size is not surprising when viewed against the economies of scale in refining. Placing operating costs per unit intake in a refinery with a capacity of 1 million tons a year at 100, costs for a 2 million ton refinery drop to 75 and for a 5 million ton refinery to about 55. Together with the growth in the size of refineries, there has been a marked trend towards the construction of larger and larger individual distillation units. The largest ones now have a capacity of nearly 10 million tons a year. In future 12.5 million tons units can be considered realistic.

222.   The growth in the size of refineries implies a considerable growth in the required site areas. The determining factor is not the space needed for the processing units, but rather the room required for the storage of crude oil and finished products, not only for commercial reasons but also to fulfill the security stockpiling obligations (see Chapter IV, C). Moreover, the selection of suitable sites will become more difficult because of the increasing concern over the effects of refinery operations upon the environment. For these reasons it will become more and more difficult to find sites of the required size and in the desired locations. However, in view of economics continuing to favour the import of crude rather than of refined products, it may be expected that this problem will be resolved satisfactorily. Everything points to the need for governments to undertake more long range planning of industrial areas, where future plants of the oil, chemical and heavy industries can be concentrated.

THE PRODUCT PATTERN OF REFINERY OUTPUT

223.   Table 22 shows the trend in refinery output from 1960 to 1970. The most striking developments are the declining share of motor gasoline and the increasing share of gas/diesel-oil in total refinery output. Gas/diesel-oil owes its increasing share to the very rapid growth in the domestic sector (central heating). The output of heavy fuel has shown a fairly steady trend in percentage terms, during a period when the crude supply (see Chapter III, B) has been increasingly in terms of the lighter crudes from North and West Africa.

224.   There has been a striking growth in demand for feedstocks for the petrochemical industry, in particular for naphtha for the manufacture of olefines and aromatics. In 1960 the deliveries of feedstocks from the oil industry to the petrochemical industry accounted for 2 per cent of total petroleum consumption, increasing to more than 9 per cent by 1970. At individual locations where the petrochemical industry is concentrated this percentage is of course considerably higher. The demand for petrochemical feedstocks has assisted refiners to balance their output with the lighter crude stream referred to above.

96

TRADE IN REFINED PRODUCTS

225.   In 1970 the total export of refined products from the OECD European area (18.4 million tons) was equal to about half the quantity of products imported into the area (37.6 million tons).   If bunkers were counted as exports, the area would have been a net exporter that year.

226.   The volume of trade in oil products between OECD Member countries in Europe more than doubled between 1960 and 1970, and by 1970 greatly exceeded the product movement to countries outside Europe (in which year trade between the OECD European countries amounted to 103.1 million tons compared with exports of 18.4 million tons to countries outside the OECD European area).

227.   In 1970 the Netherlands were the main source of product exports, closely followed by Italy and then by the UK, France and Belgium.   The main product importing countries who by 1970 were more dependent on imports than on their own refineries for their total product supply were Denmark, Sweden and Switzerland.   The products imported from outside the OECD European area in 1970 were mainly gas/diesel-oil and residual fuel-oil (together accounting for 80 per cent of total imports).   57 per cent of the gas/diesel-oil and fuel-oil imports from outside the OECD European area came from Eastern European countries.

DEVELOPMENTS IN PRODUCTS QUALITY

228.   Motor gasoline octane ratings showed little change during the period 1960 and 1970 with respect to what are generally termed the premium grades, but the so-called regular gasoline octane numbers were generally raised.   Octane ratings typical for 1960 were 97/86; by 1970 99/91.

229.   Current and proposed legislation can affect gasoline composition and properties in two ways; firstly, directly from legislation which limits lead content, and secondly, indirectly, from exhaust emission legislation which itself could lead to a substantial reduction in the use of lead in motor gasoline.   For example, Germany has legislated for a reduction "across the board" to 0.4 gm lead/litre by 1972 and to 0.15 gm/l by 1976.   Sweden has proposed that one grade should be unleaded by 1974, and there is an edict that all gasoline should be unleaded by 1980.   Environmental problems due to production, refining and consumption of oil are discussed in detail in Chapter VII.

230.   The full sphere of legislation regarding lead in gasoline is still unknown and the extent and timing of restrictions cannot be forecast. Refiners are, however, likely to be faced with the need to invest in upgrading plant to compensate for the partial or total withdrawal of lead; the extent of the investment depends on the legislated requirements and on engine design/octane requirements.

231.   The extent to which the refining industry in Europe will be required to invest in new facilities to meet more stringent sulphur content require-

ments will depend on the volume and regional concentration of fuel-oil demand, its "sulphur spectrum", the availability of low sulphur crude oil and of course on how stringent the requirements will be.

C. REFINERY DEVELOPMENT IN NORTH AMERICA

232.   Refinery capacity in the United States has grown on comparative modest terms, at 3.5 to 4 per cent per year from 1960 to 1970, as Table 21 shows.   Yet this expansion was more than sufficient to meet the demand requirements of at least the first half of the decade, to the extent that except for 1966-69 the oil refining industry has been operating below optimum utilization.   The expectation is that growth will continue to be modest during the 1970s, somewhat below the anticipated growth rate in demand for refined products, which is estimated at 3.8 per cent per year.   A substantial portion of the 1970-80 demand growth will occur in the residual fuel market, where most of the supply will be imported.   Thus, additions to distillation capacity will be moderately lower than during the preceding ten years, with principal emphasis focused upon capacity additions to manufacture gasoline and commercial jet fuel.   The construction of new conversion and upgrading capacity to increase the yields of the more valuable lighter products has been considerably more rapid, particularly with respect to hydrocracking and catalytic reforming facilities.   Refinery capacity in Canada has shown a growth rate similar to that of the United States during the period 1960 to 1970, with a somewhat less pronounced growth in secondary plant capacity.

233.   In the United States and Canada there has been a noticeable trend to the closure of small, old and uneconomic refineries and a consolidation into a smaller number of larger plants.   For example, by 1970 over 50 per cent of the refinery capacity in the United States was concentrated in 12 per cent of the total number of refineries, with an average capacity of about 10 million tons each.   Still, average intake capacity in North America is much lower than in OECD Europe and Japan (Table 24).   Increasing competitive pressures are likely to lead to further concentration and to increases in the average size of refineries in the future.

234.   With respect to refinery output, the yield of gas/diesel-oil and residual fuel-oil has remained essentially constant in percentage terms.   In 1970 residual fuel represented about 8 per cent of the refinery output, which is in sharp contrast to the European yield of 40 per cent, shown in Table 22.   Apart from the fact that industry in North America uses coal and natural gas to a much larger extent than does industry in Europe and Japan, the unrestricted import of residual fuel-oil into the United States obviates the need for domestic manufacture of this product on a mass scale.   Moreover, residual market prices historically have been below the price of domestic crude, particularly on the East Coast and some point to this disparity as a factor in the decline of heavy fuel refining yields.

235.   The growth in the US demand for residual fuel-oil during 1968-69 and again in 1970 was particularly marked, being the consequence of increased public utility demand, delays in nuclear plant construction, limited

availability of natural gas and the switch from coal to low sulphur oil for pollution reasons. Imports that year covered 75 per cent of total inland consumption of residual fuel. The main area of consumption of this product is the East Coast.

236.   Substantial increases in the price of imported fuel-oil, together with a wider recognition of the increasing dependence of the US on imported crude oil in the future, and of the vulnerabilities implicit in such dependence may be expected to lead to a reversal to some extent in the pattern of refining in the US with emphasis in the years ahead on increased domestic production of residual fuel.

237.   In the US the petrochemical industry— in particular that part of it based on ethylene and its derivatives— has depended largely on natural gas for its feedstock. US natural gas is relatively low-priced and of the proper chemical composition— in contrast to European gas— for use as a petro-chemical raw material. As a result this industry has been less dependent on the oil refining industry for its feedstock supply, than Europe or Japan. However, with the anticipated growing shortage of natural gas in the US, the petrochemical industry out of necessity may be forced to look more to refining feedstocks, such as naphtha.

238.   Motor gasoline octane numbers in the United States and Canada showed far less movement during the period 1960 to 1970 than in the years prior to 1960. For example, typical octane numbers of the three United States grades of 102/99/89 in 1960 moved to 103/100/94 by 1970. In Canada the two grades 99/92 were raised to 100/95 by 1970.

239.   In the US, the Environmental Protection Agency (EPA) has given notice that it plans to propose regulations for the lead content of motor gasoline along these lines: to require one grade of non-leaded gasoline by 1974, and all grades to contain not more than 0.13 gm/litre "as soon as possible". The ultimate intention is to require all grades to be unleaded.

240.   In the area of pollution of the environment by sulphur compounds deriving from the combustion of fossil fuels, the normal practice has been for the US Federal Government (through the Environmental Protection Agency) to set recommended standards; authorities in states and cities then establish local regulations by applying these standards as considered appropriate. Present local legislation varies greatly throughout the country and is in a constant state of flux. To illustrate, fuel-oil sulphur limits are subject to broad geographic interpretation:

New York City Metropolitan . . . . . .   initially a maximum 1 per cent sulphur.
                                                              reducing on 1/10/71 to 0.3 per cent sulphur (except for utilities built prior to 20/5/67 which continue at 1 per cent).
New York State . . . .   3 per cent sulphur maximum.
New Jersey . . . .   initially 0.5 per cent sulphur maximum reducing to 0.3 per cent from 1/10/71.

The move to set strict sulphur content regulations has led to a rapid growth in investments in desulphurizing units in the Caribbean area, the traditional supplier of the US for fuel-oil. European suppliers of low sulphur fuel-oils have been attracted to the US market in recent years.

D. REFINERY DEVELOPMENT IN JAPAN

241.   As illustrated in Table 21 Japan has become the largest refining country in the OECD after the US. During the past five years the growth rate has been in excess of 13 per cent per year.

242.   Expansion of refinery capacity is controlled by MITI (the Ministry of International Trade and Industry). The demand for naphtha (as a petrochemical feedstock) and for fuel-oil is growing, faster than for other products, and because the level of refinery capacity is geared to the demand for middle distillates there is a substantial deficit at the top and bottom of the barrel which has to be met by imports.

243.   The pressure of population upon the limited area of habitable land is making the provision of sites for extension of refinery capacity increasingly difficult. Reclamation of land at coastal sites is becoming normal practice, and refineries will more and more be located away from the main centres of population. In connection with this location problem it may be mentioned that a special group of experts in oil refining investigated the feasibility of the "oil refining tanker" method and the "pontoon refining" (a large platform in the sea) method, mainly from the technological and economic aspects. The interim report concluded that these methods are hardly feasible at this stage (though it may be worthwhile considering them as future possibilities) for the following reasons:
   1.   the doubtful safety of refining on a tanker or pontoon; and
   2.   the very high costs compared with a normal refinery on the coast, even taking into account the rapid rise in land values.

244.   In terms of petrochemicals, Japan has been the fastest growing country in the world. Feedstocks to petrochemicals (mainly naphthas) grew from about 3 per cent to 12 per cent of total petroleum consumption in Japan in the 1960s.

245.   As a consequence of the rapid industrial development, and the concentration of industry in congested areas of Japan, pollution in all its forms has become a major cause for concern. Within the framework of the law governing control of air pollution, sulphur levels in fuel-oil and emission of sulphur dioxide are under regulation, as are lead levels in motor gasoline.

246.   In 1970 a maximum lead content in motor gasoline of 0.3 gm/ litre became a legislative requirement. MITI are proposing to require total removal of lead by 1974, but are considering partial relaxation depending on the requirements of the car population.

247.   Legislation covering atmospheric pollution in Japan defines three low sulphur areas: congested areas, polluted areas and preventative measure

areas. The local regulations are in the hands of the Prefectures, and limitations are typically expressed in terms of ground level sulphur dioxide concentrations, total sulphur emissions and sulphur content of the fuel oil burned. Examples of typical target figures for sulphur content are set out in the following (per cent):

	1973
Congested areas	0.9
Polluted areas	1.3
Preventative measure areas	1.5

248.   Another aspect of national policy affecting refiners is the availability of fiscal incentives for the import of low sulphur crude oils and for investment in hydrodesulphurizing facilities.

E. REFINERY DEVELOPMENT IN AUSTRALIA

249.   Prior to 1954 there were only three small refineries in Australia. Now there are ten modern refineries with a total crude oil distillation capacity of 30.5 million tons per year. All refineries have substantial catalytic reforming capacity and all except three have catalytic cracking units. The first hydrocracking plant in Australia was commissioned in 1967. Three refineries produce lubricating oils and seven are equipped to manufacture bitumen.

250.   The major change in Australian refineries has resulted from the introduction of local crude oil in large quantities. The refineries were designed to process a high proportion of either the high sulphur Middle East or the waxy South East Asian crudes. The discovery of major oil fields in Bass Strait and Government policy aimed at ensuring that the maximum amount of indigenous crude oil will be refined locally have altered the crude input pattern considerably, especially in the last two years.

251.   The crudes so far discovered in Australia are low in sulphur, containing high percentages of the more volatile gasoline, kerosene and gas-oil fractions and very little residual. The heavier fractions of the Bass Strait crudes are waxy and have a high pour point.

252.   Modification has been necessary to existing refinery plant, particularly provision of increased catalytic reforming capacity, to enable processing of the increased volume of local crude. One refinery is now processing Australian crude only. The prospect of further Australian commercial oil fields being found is a complicating factor for long range construction programmes.

253.   In the middle 1960s refinery primary distillation capacity closely approximated total consumption of petroleum products. However, the output pattern was such that an excess of fuel-oil and diesel-oils were produced and motor spirit and aviation turbine fuel were in short supply (see Table 22). Fuel-oil became the principal petroleum product exported until this imbalance was corrected.

254.   Indigenous crude is deficient in the heavy end of the barrel while inland fuel-oil demand has increased by 66 per cent in the last five years. Indigenous crude now comprises over 60 per cent of refinery input. As a result, in 1969, Australia changed from being a net exporter to being an importer of fuel-oil.

F. FUTURE TRENDS

255.   The growth in refining capacity in the OECD countries will— other factors remaining equal— depend of course on the growth in demand for petroleum products.   The largest growth in refinery capacity can therefore be foreseen during the next ten years for Japan, followed by the European Member countries and North America.   For Europe this would imply almost a doubling of existing capacity; in Japan the increase would have to be even more dramatic: $2\frac{1}{2}$ times existing capacity.   Refinery expansion in Europe and Japan announced so far, mostly for completion during 1971-1975, come to a cumulative capacity of about 450 million tons throughput per year.   Of this, 320 million tons per year or 70 per cent of the forecast additional capacity will be located in Western Europe.

256.   Such expansion will undoubtedly place a heavy burden on both governments and industry in the planning of industrial areas, enlarging of refining capacities at existing locations, and selecting new sites.

257.   The building of refining capacity "at the source", that is, in the oil producing countries, is taking on greater promise.   Apart from possible pressures in the future from the host governments in order to stimulate industrial activity and local employment, it has been suggested that such a trend might be stimulated by two factors now becoming important in consuming countries: the problem of environmental pollution and the increasing difficulty in finding suitable sites for the refinery capacity expansion required to meet future growth.   However, the location of refineries distant from the consuming markets is unattractive from several points of view. Firstly, it remains more economic to transport crude whether by tanker or by pipeline rather than products to the consumers, and to process different types of crude with complementary qualities centrally, rather than at separate locations, which must then be followed by blending in consuming areas to maintain quality and/or by additional refinery investment to deliver all products on specification.   Secondly, as illustrated in Chapter VI, economies of scale in tanker transport have found more immediate application in the movement of crude oil.   Crude oil transport is increasingly being made in vessels of 250,000 dwt or larger.   The largest vessels currently envisaged for products are of the 100,000 dwt class.   This means that in the already crowded shipping lanes— in particular the English Channel— if future growth were to be partially met by product rather than by crude import, the number of tankers would increase by a factor of 3 or more.

258.   The shift in Europe from coastal to inland refineries appears to have run its course.   No further stress on inland construction is expected in the foreseeable future.

259.   Refinery construction and operating costs are likely to increase, the extent of this increase depending largely on inflation and requirements set by considerations for preservation of the environment.   Savings obtainable by economies of scale and improvements in technology and efficiency as evident in the past decade, no longer seem adequate to compensate for inflationary cost increases, the latest of which may be seen by the following Nelson index[1] figures:

REFINERY CONSTRUCTION (1946 = 100)

Nelson index

	1954	1960	1967	1968	1969	1970
Refinery inflation index	179. 8	228. 2	286. 7	304. 1	329. 0	364. 9
Construction and Design	1. 708	2. 211	2. 767	2. 816	3. 092	3. 092
Productivity Construction (true cost) index	105. 3	103. 2	103. 6	108. 0	106. 4	118. 0

260.   Social developments are likely to have a major influence on the refining industry and on product costs.   Low lead gasolines and low sulphur fuels are prime examples of fuels for which efforts to meet pollution abatement standards are yet to be reflected in higher costs and in higher prices.

261.   For gasoline, the extent of the investment required in a new refinery plant, to compensate for reduction in lead content, will depend on the allowable lead levels, the octane requirements of the market and the precise nature of the exhaust emission legislation.

262.   The cost of manufacturing low sulphur fuels will make itself increasingly evident in Europe.   The installation of desulphurizing units, which involve a cost ranging currently from 30 cents to 50 cents per barrel of oil per one per cent of sulphur removed, will be the major contributing factor.

---

1. The Nelson indices, which are published monthly in the Oil and Gas Journal, relate to US data, but it is believed that the trend is also valid for other parts of the world.   Those indices show the movement since 1946 in the cost of refinery and processing construction based on US figures.   They include:
   a)   An Inflation Index showing the cost of duplicating a refinery installation at dates following the base year of 1946 without regard to changes in mechanical or process design or construction techniques.
   b)   A Construction and Design Productivity factor taking into account the improvements that have been achieved in construction and process design.
   c)   A True Cost Index that combines the above two indices to show the relative cost of installing processing capacity.

Table 20. REFINERY CAPACITY IN THE CARIBBEAN AND THE MIDDLE EAST 1960, 1965 AND 1970

Million tons

	1960	1965	1970
Caribbean	88	133	194
Middle East	61	85	114

SOURCE: BP Statistical Review of the World Oil Industry.

Table 21. REFINERY CAPACITY IN OECD MEMBER COUNTRIES
1960, 1965, 1970 AND 1975

Million tons at end of year

	1960	1965	1970	1975
Austria	2.2	4.7	8.2	14.5
Belgium	8.6	16.7	35.9	44.5/50.5
Denmark	0.2	3.8	10.8	10.8/12.0
Finland	1.2	3.1	10.0	15.0
France	40.2	71.2	116.5	180.0
Germany (FR)	40.5	81.4	120.3	205.6
Greece	1.8	4.3	4.7	5.7
Iceland	-	-	-	-
Ireland	2.0	2.0	2.6	2.6
Italy[1]	45.7	97.9	136.5	156.9
Luxembourg	-	-	-	-
Netherlands	22.8	31.9	68.5	100.8/105.8
Norway	0.1	2.9	6.4	8.2
Portugal	1.3	1.7	3.8	5.8/11.8
Spain	11.1	16.0	34.8	53/56
Sweden	3.2	4.0	12.6	19.9
Switzerland	-	2.0	5.5	6.8
Turkey	0.4	5.2	7.5	15.1
United Kingdom	50.0	73.3	114.8	154.0/180.0
Total OECD Europe[2] ...	231.3	422.1	699.4	999.2/1,046.4
Canada	46.9	53.2	69.6	n.a.
United States	474.0	507.7	653.7	n.a.
Total OECD North America	522.1	560.9	722.3	n.a.
Japan	33.4	93.9	159.2	n.a.
Total OECD[3]	765.6	1,076.9	1,581.9	n.a.

1. Excluding Petrochemical plants using crude oil.
2. Excluding Finland.
3. Excluding Finland and Australia.
N.A. = Not Available.
SOURCE: OECD Oil Committee.

Table 22.   TRENDS IN REFINERY OUTPUT 1960-1970

Percent of Total

	1960	1965	1966	1967	1968	1969	1970
OECD Europe							
LPG	1.8	2.1	2.1	2.0	1.9	1.9	1.8
Aviation fuels	3.0	2.5	2.6	2.6	2.5	2.7	2.8
Motor gasoline	18.0	14.5	13.6	13.4	13.4	12.5	12.0
Kerosenes	2.6	1.7	1.6	1.7	1.7	1.6	1.5
Naphtas	-	-	-	4.5	5.1	5.3	5.4
Gas/diesel oil	25.7	28.2	28.3	28.1	29.2	29.5	30.1
Fuel oil	40.7	41.0	41.3	41.0	39.9	40.1	40.2
Others	8.2	10.0	10.5	6.7	6.3	6.4	6.2
Total	100	100	100	100	100	100	100
North America							
LPG	4.5	1.8	1.8	1.8	1.8	1.8	1.8
Aviation fuels	6.6	5.9	6.1	6.9	7.5	7.4	6.7
Motor gasoline	41.1	40.9	41.1	40.8	40.8	41.6	42.0
Kerosenes	3.6	2.9	2.9	2.7	2.6	2.6	2.3
Naphtas	-	-	-	0.8	0.7	0.7	0.8
Gas/diesel oil	34.4	22.8	22.4	21.9	21.7	21.3	21.9
Fuel oil		9.3	9.0	9.4	9.0	8.5	8.2
Others	9.8	16.4	16.7	15.7	15.9	16.1	16.3
Total	100	100	100	100	100	100	100
Japan							
LPG	1.9	2.6	2.3	2.2	2.2	2.1	2.1
Aviation fuels	1.5	1.6	1.6	2.0	2.1	1.9	1.3
Motor gasoline	16.5	12.0	11.2	10.4	10.4	9.8	9.6
Kerosenes	6.8	6.6	6.2	6.8	7.3	7.5	8.8
Naphtas	-	-	-	8.0	8.4	9.3	10.1
Gas/diesel oil	66.5	11.3	11.6	12.3	11.6	11.6	12.0
Fuel oil		54.7	54.8	54.5	54.0	53.9	51.7
Others	6.8	11.3	12.3	3.8	3.9	3.9	4.4
Total	100	100	100	100	100	100	100
Australia							
LPG	0.2	0.8	0.8	0.9	1.1	1.3	1.2
Aviation fuels	2.6	2.1	3.1	3.6	3.8	3.9	4.8
Motor Gasoline	32.5	30.9	30.0	30.4	29.7	30.9	30.9
Kerosenes	1.5	1.3	1.2	1.1	1.1	1.0	0.9
Naphtas	-	-	-	-	-	-	-
Gas/diesel oil	22.9	16.9	17.2	17.1	17.4	19.1	19.2
Fuel oil	29.2	31.5	31.4	30.2	30.3	26.2	26.4
Others	11.1	16.5	16.3	16.7	16.6	17.6	16.6
Total	100	100	100	100	100	100	100

NOTE: All figures are taken from OECD Oil Statistics, except the 1960 figures for North America and Japan, which are derived from OECD Statistics of Energy, and all the figures for Australia which have been provided by the Australian Delegation.

106

Table 23. OECD EUROPE REFINERY LOCATION[1]

	1959	1962	1964	1965	1966	1967	1968	1969	1970
Million tons									
Inland	29.2	56.5	107.0	119.7	139.3	162.7	184.5	193.5	209.7
Coastal	165.0	214.5	281.4	299.3	329.4	372.4	399.7	455.9	487.7
Total	194.2	271.0	388.4	419.0	468.7	535.1	584.2	649.4	697.4
Percent of total									
Inland	15.0	20.8	27.5	28.6	29.7	30.4	31.6	29.8	30.1
Coastal	85.0	79.2	72.5	71.4	70.3	69.6	68.4	70.2	69.9

1. Excluding Australia and Finland.

107

Table 24. AVERAGE INTAKE CAPACITIES OF REFINERIES
1965 AND 1970

Tons per day

	1965	1970
OECD Europe	7, 500	11, 200
Canada	3, 500	4, 500
United States	5, 300	7, 000
Japan	6, 600	12, 500
Australia	6, 000	8, 500

VI

TRANSPORT

A. SEA TRANSPORT

263.   We noted in our 1964 report "Oil Today", that oil accounted for about half of the world's seaborne freight and that the quantity carried (640 million tons in 1962) had increased by about 40 per cent in the previous three years.   The rapid growth in international oil movements continues and oil's lead as the largest single element in world trade has been consolidated, both in volume (55 per cent) and in value (10 per cent); tankers and combined carriers now represent almost half the total dead-weight tonnage of the world fleet.   OECD Member countries' imports account for about 85 per cent of the world's movements of oil by sea.

Developments following the six-day war

264.   Closure of the Suez Canal following the Arab-Israeli war in 1967 brought about major shifts in the pattern of oil movements.   Vessels moving crude oil to Europe from the Persian Gulf were faced with the long voyage around the Cape of Good Hope, a round trip of some 22,000 miles taking about 65 days.   By comparison, round-trip movements using the Canal represented 12,500 miles and required 40 days.   Maps 1 and 2 depict oil movements by sea as they stood before the Six-Day War and as they stood in 1970.   Freight rates rose steeply, reflecting the severe demands placed upon the tanker fleet.   This rise in rate is illustrated in Figure 7 showing the movements of a weighted average of the three AFRA (Average Freight Rate Assessment) quotations and the single voyage quotation during the period 1964-71.

265.   Although freight rates rose steeply in the second part of 1967, they did not reach the level attained during the 1956-57 crisis.   At that time spot rates went up to a highest point of +253 per cent[1], whereas in 1967 the peak spot rate was +91 per cent.   Apart from the differences between the two crises, this disparity in increases in tanker rates can be explained by the fortunate coincidence that at the beginning of 1967 a large reserve in tankers was available.   Thus, despite the continued closure of the Suez Canal freight rates declined after 1967 and precrisis levels were restored in the first part of 1969.   However, much of the margin between supply

---

1.   That is, spot tanker rates were this much higher than the flat rate.

and demand for tanker capacity had now been lost. This loss became evident when at the end of 1969 the demand for tanker capacity began to increase for reasons elaborated upon in Chapter III. During 1970, spot rates jumped very sharply, to triple the 1967 peak, fostered in part by the reduction in Libyan production and the shut-down of the Trans-Arabian pipe-line (TAPLINE). During 1971 they fell sharply to reach, in the third quarter, their lowest levels since the Suez crisis. In the fourth quarter they recovered slightly though, as it proved, temporarily.

TANKER SIZE AND COSTS

266.   Tables 25 and 26 show the dramatic increase in the proportion of tanker tonnage now accounted for by the Very Large Crude Carrier (VLCC): in 1961-65 the largest of the tankers (over 125,000 dwt) represented only about 0.3 per cent of the tonnage constructed; but in the following six-year period, 1966-71, the largest size tanker (205,000 dwt and over) represented 50 per cent of the tonnage constructed. As well as this marked increase in the construction of VLCCs, there was a steady development in specialised vessels.

267.   The striking increase in tanker size has coincided with a sharp rise in tanker building costs. In 1967 shipyards were quoting a price of about $125 per dwt for 30,000 dwt tankers and $100 per dwt for 70,000-tonners. By the end of 1970 these prices had increased to $300 per dwt and $190 per dwt respectively. Construction costs of a 250,000 dwt tanker had doubled over the same period ($70 per dwt to $140 per dwt). Insurance charges had risen even more steeply. Transportation can represent a very important factor in the cost (price) of crude oil to the buyer. Of the total cost of crude oil delivered in Northwest Europe out of the Persian Gulf, some 30 per cent (at Worldscale 70 out of the Persian Gulf and around the Cape) is transport cost. Because of the importance of the transport element in delivered costs of oil, the economies of scale obtained in the deployment of the larger tanker are of critical importance. A 30,000 dwt tanker carrying oil out of the Persian Gulf to Rotterdam via the Cape of Good Hope at the end of 1970 could do so at total transportation costs of $9.93 per cargo ton; a 70,000 dwt tanker could perform the same function at $7.45 per cargo ton and a 250,000 dwt tanker at just $5.44 per cargo ton. (Table 27 gives nominal Worldscale-tariffs for some important oil routes.)

268.   Average oil tanker tonnage will continue to increase over the next few years because of the growing number of VLCCs in the world fleet (tankers of over 160,000 dwt account for 87 per cent of the tonnage on order). The reason for this trend lies in the considerable economies of scale possible with the present generation of VLCCs (see Figure 8). However, increasing tanker tonnage beyond 200-300,000 t, which is now a common size, presents certain technical problems (difficulty of handling very large vessels) and economic problems (provision of the necessary harbour facilities). This could well check the escalation in oil tanker size and the average tonnage of VLCCs in the immediate future could well remain much the same as that of the tankers now being delivered. Even so, it should be noted that a number of 500,000 t ships are now being built.

COMBINED CARRIERS

269.   Normally a price has to be paid for flexibility. With tankers the lowest construction and operating costs are achieved with ships designed solely for the carriage of oil.   But operational flexibility is sometimes considered sufficiently desirable to justify higher building and operating costs. Multi-purpose bulk carriers can carry oil in one direction and dry bulk over a major part of the return voyage; if the two freighting aspects can be approximately matched geographically, substantial economies in total freight costs can be achieved as the proportion of ballast mileage versus loaded mileage is minimized.   Combination carriers now represent about 10 per cent of tanker tonnage; this percentage is expected to grow to 15 per cent or more by 1975.   As such carriers can be (and are, if charter rates warrant it) employed wholly in the oil or bulk trades, they have the additional advantage of providing a substantial cushion of extra carrying capacity should the need arise.

PRODUCTS CARRIERS

270.   Though most interest focuses on the VLCC, smaller vessels continue to form an essential part of the oil industry transport system.   Seaborne trade in products in fact accounts for about one-fifth of total oil tonnage moved by ships and has now reached an estimated 220 million tons per year. The world's oil products carrying fleet— tankers of about 30,000 dwt are at present generally considered the upper limit for the products carriers— includes some 1,700 vessels with total capacity well over 32 million dwt. Some of these ships, particularly those in the 25,000 to 30,000 dwt class, continue to be used for carrying crude oil, but the vast majority are or can be used as products carriers.

271.   Many of these vessels were originally built as crude oil carriers and were later converted to the more specialized trade; some still switch back and forth from products to crude or from white oils to black oils, depending on demand for their services.   But the cleaning necessary to make a dirty tanker ready for the carriage of white oils is expensive and not always practical.

272.   A few ships were built specially as multi-cargo tankers early in the 1930s but it was not until the post-war era that a significant number of ships specifically designated to carry products came into use.   Those now being commissioned are highly sophisticated with some vessels capable of carrying up to 40 separate grades of products.   As with the VLCCs construction costs have roughly doubled in recent years; a 28-30,000-tonner now costs well in excess of $20 million[1] to build.   At the same time, the size of vessels is also going up; over 90 per cent of all products carriers are in the 20 to 30,000 dwt class.

273.   The most significant characteristic of the world's products carrier fleet is however its advancing age.   Over half the present fleet is at least 16 years old and about a third were built more than 20 years ago.   While

---

1.   The Norwegian Delegation estimates a cost of $10-12 million.

many of these ships can be expected to continue in the black oil trades, the life of a multi-grade tanker carrying products of high-quality specifications is about 15 years; this means that a growing proportion of the older vessels will drop out of white oil trading in the next few years. As in recent years most new vessels have been crude carriers— the high demand for oil combined with the closure of the Suez Canal has forced the oil industry to concentrate its attention and available investment capital on ever-bigger VLCCs at the expense of small ships to the extent that there has been relative neglect of products carriers. There are signs that normal corrective forces are at work and several companies have embarked on building programmes for products carriers. Encouragement has been found in the opinion that freight rates for this class of vessel will remain firm: though there is competition from pipe-lines for white oils many smaller markets will continue to be served most economically by sea transport; and for heavy products like fuel-oil there is often no practical alternative to the products carrier.

CARRIERS OF LIQUEFIED NATURAL GAS

274.   Natural gas, the main constituent of which is methane, becomes a liquid when refrigerated to —161 °C (—258 °F) at atmospheric pressure, at approximately 1/600th of its gaseous volume. In such a state it is well suited to transport by specialised ocean tanker.

275.   The first seaborne shipment of Liquefied Natural Gas (LNG) took place as recently as 1959, when a converted dry cargo vessel crossed from Lake Charles, Louisiana to the UK Gas Council's storage terminal at Canvey Island.

276.   Subsequently progress has been quite dramatic. The first purpose-built LNG carriers, designed to carry 12,000 t cargoes of LNG from Algeria to the UK and France, had the dimensions of conventional oil tankers of 28,000 dwt. Since then progressively larger vessels have been built to move LNG from Libya to Southern Europe and from Alaska to Japan. Deliveries from Brunei to Japan, which are planned to begin in the winter of 1972/1973, will be made in vessels of about the size of 80,000 dwt oil tankers. Some large LNG carriers on order or under construction are equivalent to 150,000 dwt oil tankers.

277.   As about a third of the world's hydrocarbon reserves are in the form of gas, located in areas often remote from the main centres of consumption there is every likelihood this specialised trade will continue to grow rapidly in importance, and will develop on international lines in the same way as the petroleum trade, although on a lesser scale. Some forecasts suggest that the number of vessels in this trade will increase 10-fold during the 1970s from 10 in 1970 to 100 in 1980, with carrying capacity rising some 25 times.

278.   Because of their complexity, construction and operating costs will always be higher for LNG carriers than for conventional oil tankers; technical developments and economies of scale have already reduced the differential and further steady progress along these lines may be expected to continue. To illustrate the difference in transport costs between LNG

and oil it may be mentioned that the construction costs for a tanker designed to carry 70,000 t. of LNG are $60 million compared with $14 million
for a 70,000 t. oil tanker. As more sources of gas supply and more LNG
carriers become available, greater flexibility of marketing arrangements and
increased security of supply will result; consumers will no longer be wholly
dependent on a particular supply source.

OWNERSHIP

279. Table 28 shows that at the end of 1971 32 per cent of the world
tanker fleet was owned by oil companies, 63 per cent by independent tanker
owners and the balance by governments. In recent years the distribution
of the world fleet between oil company and private ownership has remained
roughly constant though there have been year-to-year variations.

280. Some 60 per cent of the fleet at the end of 1971 was under Libe-
rian, British, Norwegian and Japanese flag; "flags of convenience", which
companies adopt primarily for tax reasons, account for about 30 per cent
of the total tonnage.

POLLUTION

281. Concern over pollution has increased markedly in recent years;
Chapter VII deals with the oil aspects of pollution, including those relating
to sea transport. Here it will suffice to say that although steps to reduce
and, if possible, eliminate pollution are often desirable in themselves, it is
rare for such measures not to add to costs of operation. Thus the industry's Load on Top (LOT) system under which residues are retained on board
during the washing process rather than discharged into the sea has been
estimated to add some 6 cents a ton to carrying and processing costs. The
compensation provisions of TOVALOP (Tanker Owners Voluntary Agreement Regarding Liability for Oil Pollution) and CRISTAL (Contract Regarding an Interim Supplement to Tanker Liability for Oil Pollution), are
plainly desirable, but must also be considered as additional elements in
transport costs. In addition, certain nations have legislation under which
taxes for purposes of compensation for pollution are imposed.

FUTURE PATTERNS

282. Total size of the tanker and combination carrier fleet has developed
as shown below:

				Million deadweight tons.
	1960	1965	1970	1975 (est.)
Tanker tonnage	60	80	140	250
Combined Carriers	1	3	14	40

283. At present about 75 per cent of the combined carrying tonnage
is trading in oil. Assuming a similar proportion for 1975 the total oil
carrying capacity will be 280 m. dwt against 150 m. dwt in 1970, an increase

of 86 per cent, equivalent to an average annual increase of 13.3 per cent. Currently the growth of tanker tonnage demand is not as high as this. But an increasing part of the big US market will have to be supplied from sources involving tanker transport for at least part of the journey; there is no present prospect that discoveries in the European Continental Shelf will significantly alter this picture.

284.   It is impossible to forecast the future pattern of freight rates. In the past the market has been both complex and volatile, and there is no immediate prospect of change. Nevertheless it would be surprising if long-term levels settled down to the rates obtaining before the Arab-Israeli War of 1967; both construction and operating costs have risen steeply for all sizes of vessel and further significant improvements in productivity appear unlikely.

B. PIPELINES[1]

285.   Pipelines have been used for transporting oil since the first years of commercial crude oil production. Where either production or main areas of consumption are some distance from the sea there is indeed no practical alternative to pipeline transport, and towards the end of the 19th Century the Russians had built crude oil pipelines from the Baku oil fields to the Black Sea; one was over 400 miles (650 km) long. But it was in the United States, which till recent years both produced and consumed well over half the world total of petroleum, that the modern pipeline industry developed. Since World War II— and especially in the past decade— a number of important pipeline systems have however been built in other areas. In general product pipelines have developed less rapidly than crude lines because economic alternative transport systems have been available; but here also steady progress has been made in recent years, in the United States and elsewhere.

CRUDE OIL PIPELINES

286.   Most of the cost of pipeline transport arises from capital charges and fixed operating costs, the former being the main component. Pipelines are therefore generally economic only at full and constant throughput; they can compete with tankers only if the pipeline route is considerably shorter than the tanker route, if sea transport is subject to exceptional charges such as heavy canal or port dues, or if the port nearest the centre of consumption is limited to accepting tankers uneconomically small in size; a rough guide for many years has been that a pipeline is not worth constructing to replace a sea route unless it halves the distance involved. While the use of large diameter pipe in the pipelines built in recent years has reduced considerably pipeline transport costs per ton mile, these economies have been matched till recently by the growth of VLCCs, though the rise of tanker costs has now become markedly greater than that of pipeline costs.

---

1.   For more detailed information see also "Pipelines in the United States and Europe and their Legal and Regulatory Aspects", an OECD publication prepared by the Special Committee for Oil, Paris, 1969.

287.   In a producing field the lines from the wellhead to the gathering centre in themselves often constitute pipelines of substantial diameter and length.   The trunk oil pipeline carries production from the gathering centre to the nearest refinery or ocean terminal.   Table 30 lists the more important crude lines in the world.   Other important crude lines are planned, including the 1,300-kilometer (800-mile), 120 cm. (48-inch), Alyeska pipeline which will transport crude from Prudhoe Bay in North Alaska to Valdez in the South.   Construction has been delayed because of amenity objections (there is fear of damage to the permafrost).   Contracts have been signed for the construction of two 106 cm. (42-inch) lines for Egypt's Suez-Mediterranean (Sumed) pipeline.   No decision has been reached on the projected 1,700 km. long, 106 cm. (42-inch) line from Iran (Ahwaz) to Turkey (Iskenderun) although a study has been completed.   A proposal for a line from Southern Iraq to Tartous in Syria is in the preliminary planning stages.   In addition there are plans for the extensive "looping" of some major crude lines, e.g. the NWO line to the Ruhr and the SEPC line to the Upper Rhine.

288.   In the Soviet Union, a number of extremely long, large diameter (122 cm.) crude oil pipelines are being built or planned to move West Siberian crude oil both to the East and West.

289.   Construction of new trunk pipelines might be expected to improve initially the oil industry's transport flexibility.   But as hire contracts are long term this flexibility is limited.   Moreover pipelines traversing politically unstable areas are vulnerable to disruptive action.

PRODUCT PIPELINES

290.   Product pipelines have been well established in the United States for many years; the 105,000 km. (65,000 miles) of product pipelines in the United States roughly equal the total railway mileage in the country.   The Colonial pipeline of 76/90 cm. (30/36 inches) diameter is 5,000 km. (3,110 miles) long and is by far the world's largest oil products line.   Some three-fifths of all bulk products made at US refineries are moved out by pipeline.

291.   As with crude lines, product pipelines have developed much more slowly in Europe than in the US.   The first European products pipeline was a hot fuel line built in 1917 from Glasgow to Grangemouth in Scotland to supply the British Navy with bunkers; but the first commercial line may be taken to be the TRAPIL line which started operation early in 1953 and now supplies products from refineries in the lower Seine region to the Paris area.   Important products pipelines are also listed in Table 30.

292.   For the future as oil demand in Europe increases, product pipelines from large and flexible coastal or inland refineries to centres of consumption are likely to grow in importance.   Multi-product pipelines generally need a minimum throughput of about 20,000 b/d (1 million tons/year) to compete with other forms of transport.   But because distances by rail, road and river are usually greater than by pipeline because loading and unloading operations add to the cost, pipelines may compete at throughputs lower than indicated.   Product lines may be of considerable value socially in reducing

road transport congestion; also, in some circumstances— e.g. supplies to air fields— there is the advantage of eliminating opportunities for contamination. Even so, movements of oil by conventional methods, i.e. rail, road and river, will continue on a large scale because they are more flexible and economically attractive for the movement of smaller quantities; they are also extensively used for viscous products such as heavy fuel-oils.

293. Figures 9 and 10 show how unit costs vary with size of pipe and throughput. Figure 10 in particular amply illustrates the economics of line size diameter which explains why most long-distance cross-country pipelines are built as joint ventures rather than by individual companies each owing its own small line. A pipeline has only one feature of value: its ability to continuously transport oil. It has been the practice of financiers to lend money for construction of pipelines only after responsible shippers have "guaranteed" that the pipeline would be used for an extended period of time. The practice of the shipper furnishing a "long term throughput agreement" accounts for most pipelines being built and owned by the oil companies. This agreement usually takes the form of a commitment to "ship or pay" on a specified minimum throughput for each shipper sufficient to service the loan and to cover operating costs.

294. The development of crude oil and petroleum product pipeline systems in Japan has been obviated by the siting of refineries on the coast and by virtue of the concentration of the consuming centres along the flat coastal regions. Further inland, the demand for oil has not been sufficient to warrant pipeline construction. Moreover, land ownership in Japan is so complicated that the acquisition of rights-of-way for pipelines is very difficult. Nevertheless, it is anticipated that in the coming years oil consuming areas will emerge in the inland districts sufficient in size to support the building of oil pipelines. At the same time, in other areas the volume of oil movement by tank truck and tank car is nearing its limit and the burden must be shifted to pipeline. Recently two product pipeline projects have been planned in the Kanto area, which will be completed by the autumn of 1974. One of them, 110 km. long, 16-inch line, will transport 7 million kilolitres a year from Kawasaki, where refineries are located on the coast, to the inland area. The other is a project for a capacity of 13 million kilolitres a year with a 290 km. long, 18-inch line, starting at Chiba, another refinery area in the surroundings of Tokyo. There are also other projects for product pipelines from refineries on the coast to inland areas in various parts of Japan such as Kansai, Hokkaido, Nagoya and Kyushu.

GAS PIPELINES

295. Pipeline transport has been the only practicable way of moving natural gas to market till the advent of the refrigerated methane tanker; the latter is used only for ocean movements of gas. Compared with oil lines, gas pipelines are relatively expensive to operate because their energy content per unit volume moved is far lower: on a thermal basis a gas pipeline of a given size can carry only about one quarter as much as a similar pipeline carrying oil; and compressor stations are more costly to install and absorb more power than pumping stations on oil lines. Total costs also depend on the initial pressure of the gas at source and on whether it is discharged

into atmospheric or high pressure storage at the receiving end; even under favourable circumstances transport charges are likely to account for a large part of the total cost of supplying natural gas. In addition to purely natural gas lines, there are a number of specialized "gas" lines; these carry liquid products such as LPG and ethylene. Table 30 lists some of the more important gas lines of all types.

C. HARBOUR FACILITIES

296. The obvious factors to be considered in assessing harbour facilities are those directly affecting the loading and discharge of the tankers involved— depth, jetties, pumps, tankage, and the like. There are, however, other factors of importance, especially for reception centres— the availability of sites for storage, refining and for pipeline terminals, and of services for shipping.

297. The development of harbour facilities is necessitated by the growth in imports and the increase in tanker size. It remains to be seen if, while the rise in imports might simply require extension of existing facilities, the increase in tanker size will enforce more radical changes.

298. During the past decade the amount of oil carried at sea has grown considerably, not only in absolute terms but also relative to total world oil consumption. In 1970 the quantity of oil supplied from overseas sources was almost three times that of 1960 —about 1,250 million tons against about 450 million tons— whilst world oil consumption only doubled in the same period, from 1,015 to 2,160 million tons. An ever-increasing part of the oil consumed in the world is thus supplied from overseas, 55 per cent in 1970 against 42 per cent in 1960. For the future it can again be emphasized, as stated in Part A of this chapter, that the demand for tanker transport will rise more rapidly than the demand for oil.

299. We have noted the remarkable increase in average tanker size that has taken place since our 1964 report. Though probably at a slackening rate, there will be no doubt a further increase. Apart from economies of scale which stimulate the use of larger vessels, an increase in tanker size can prevent, given the increase of the oil quantities to be shipped, congestion in sea-lanes and harbour approaches. Given that a certain amount of oil is going to move by sea, reducing the number of carrying units through rapid growth in the average-size tanker tends to reduce both the risk of collision and grounding. While a large vessel may pose a greater individual pollution threat, a reduction in the number of incidents on the high seas or in habours is more important. Various organisations, principally the Inter-Governmental Maritime Consultative Organisation (IMCO), have conducted studies to determine whether the use of larger tankers in itself poses an increased pollution hazard. The general conclusion is that they do not.

300. The extension of harbour facilities in the past years has not everywhere followed the development of tankers. The first tankers in the 200,000 dwt class were ordered and built in the latter part of the 1960s. But many harbours were not deep enough to accommodate these tankers,

and as the deepening of harbours requires in most cases extensive investments, harbour authorities first wanted to know if a sufficient number of 200,000 dwt class tankers to justify these investments would be built.

301.   In this connection it should be noted that whereas the investment costs per ton of oil transported in a larger tanker are lower than for a smaller one, the investment costs incurred by each further deepening of a harbour tend to increase per ton of oil unloaded.   The reason is that deepening is technically linked to a broadening of the harbour, so that deepening of a harbour may be coupled with an increase in harbour dues for the larger tankers.

302.   The problem of the accessibility mentioned in paragraph 298 is confined mainly to delivery.   In the loading harbours there were relatively few problems; the major oil exporting countries are geographically very favourably situated in this respect.   In the Persian Gulf, the Eastern Mediterranean, North Africa and Nigeria there is sufficient deep water available and also sufficient room and favourable weather conditions for the single buoy mooring system.   In the Caribbean area and Indonesia, however, the depth of water does not admit the general use of supertankers.   But from these areas the oil has to move over comparatively short distances— to the United States and Japan respectively— and very large tankers are not required to load there.

303.   When in the course of 1968 the first tankers of the 200,000 dwt class were placed in service, as noted there were very few ports of discharge able to receive these vessels.   The general absence of deepwater ports has been resolved in one of the following ways:

a) Transshipment: this system involves a major oil terminal, and depot where mammoth tankers can unload their cargo, the oil being then distributed to the different refineries in smaller vessels. Transshipment depots have to be located not far away from the direct route between loading and final discharge terminals, and reasonably near to the latter.   Bantry Bay was the first exemple of such a transit depot in a consuming area.   It came into service in 1968.   Japan is favoured by good harbour conditions and 200,000 dwt tankers can be accommodated at a number of sites. But the sheer volume of crude oil imports, which make up 99 per cent of total crude supply, makes implicit the need for increasing storage capacity.   Yet the expansion of storage at existing refineries is generally precluded for environmental reasons.   As an alternative, Japan is contemplating the construction of large-scale crude terminal stations such as that now in operation in Bantry Bay.   One such station, the world's largest, is located at Kiiri; it can handle tankers up to 372,000 dwt.

b) Two-port discharge: a first harbour which can receive full-loaded mammoth tankers, and take part of the cargo, and a second harbour where the remaining oil can be discharged.   Two-port discharge may become an important factor in the handling of the VLCCs.

c) Single buoy mooring system: this system involves a buoy, floating at sea as near to a harbour as possible, whence the oil can

118

be piped directly to land. This system is much affected by local weather and stream conditions and is therefore in most cases not practicable in the Atlantic and in the North Sea. Nevertheless, single buoy mooring systems have been installed and are functioning at St. John, New Brunswick, and at the Ekofisk field in the North Sea.

d) Lifting at sea (lightening): on the high sea, as far as possible in sheltered places in the neighbourhood of the ultimate destination, part of the cargo is pumped from the larger tanker into a smaller one.

e) Use of tankers with a relatively small draught compared to the dwt capacity. In this field a study, made in the Netherlands, may be of interest. The study concerns the technical, nautical and economic features of a so-called Restricted Draught Tanker, to have a carrying capacity of 425,000 dwt and a draught of 72 feet, to be suitable for operations in the North Sea.

304.   Although most of the methods enumerated in paragraph 303 have proved feasible, in general they do not solve the problem. In most cases they require a tighter and inflexible supply pattern or a longer discharge time than does direct transport from the loading harbour to the ultimate destination harbour in the consuming area. From the point of view of oil transport, this approach is generally considered the most attractive one.

305.   It is not surprising therefore that since 1968 a number of important harbours, essentially in Europe and Japan, have been made accessible for tankers of 200,000 dwt and larger, as can be seen in Table 31. If all plans are realized, some of these harbours will be able to receive 300,000 dwt and the 500,000 dwt tankers which may be launched in the future. There is, however, a limit to the deepening of certain harbours because of the relative shallowness of the adjacent sea. For other harbours the investment costs of deepening could well turn out to be prohibitive. In these cases one of the solutions mentioned in paragraph 303 has to be applied. It can be expected, however, that the bulk of the oil supply will be directed to those relatively few harbours which are accessible for the largest tankers.

306.   The expansion of facilities for large tankers in shallow harbours adjacent to relatively deep water can be effected by:

a) deepening the harbour itself;

b) building an artificial harbour in the sea, with pipeline connections to the shore.

It depends on the prevailing circumstances which of these two solutions is the less expensive. As concerns the first solution, at a certain stage, which of course differs for each individual harbour, deepening of the harbour itself is not sufficient and consequently a channel in the sea has to be dredged. This is particularly true for North Western Europe, and the dredging requires enormous investment costs. An artificial harbour in the sea must be located not too far from the coast, otherwise pumping the oil to the coast may give rise to difficulties reflecting the low temperature of the sea. This problem can be solved by building storage capacity in the sea, but this again is very expensive.

307.　　In paragraph 296 we mentioned the question of storage at harbours. Often there is no proportional relationship between the need for additional storage and the growth in the quantity of oil supplied. When the same quantity of oil is supplied in larger and thus fewer tankers, the need for storage will grow. For instance, a doubling of the tanker size from 150,000 dwt to 300,000 dwt means that the capacity has to increase by about 40 per cent. There will then be a fairly high idle storage capacity in the interval between the arrival of successive tankers. Conversely, if more oil is supplied with the same size tankers, the storage capacity will grow less than proportionally, as idle capacity will diminish.

308.　　The time-lag between tanker and harbour developments, which we have seen in the last few years, will not necessarily recurr in the future, or at least not to the same extent. It can be expected that the development in tankers will be fairly steady, whereas harbour expansion will be more spasmodic.

Map 1

INTERNATIONAL FLOW OF PETROLEUM 1967 [1]

(In thousand barrels per day) (Estimated)

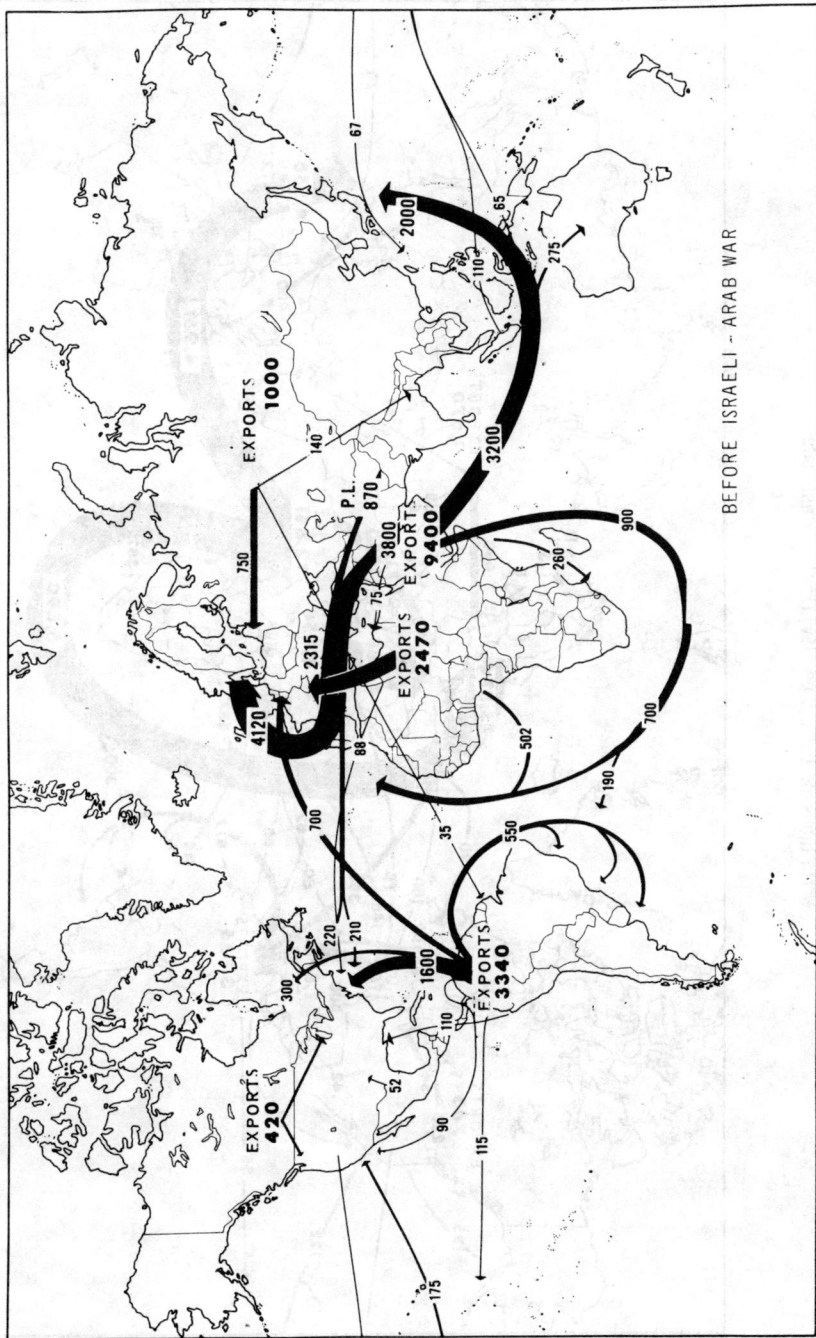

EXPORTS 1000

EXPORTS 420

EXPORTS 9400

P.I. 870

EXPORTS 3800

EXPORTS 2470

EXPORTS 3340

2315

4120

2000

3200

900

700

190

502

260

550

35

700

88

75

140

750

220

210

300

52

90

115

175

110

67

65

110

275

1600

BEFORE ISRAELI – ARAB WAR

1. Excluding imports into the USSR and other Eastern European countries, and the People's Republic of China.

121

Map 2

INTERNATIONAL FLOW OF PETROLEUM 1970[1]

(In thousand barrels per day) (Partially estimated)

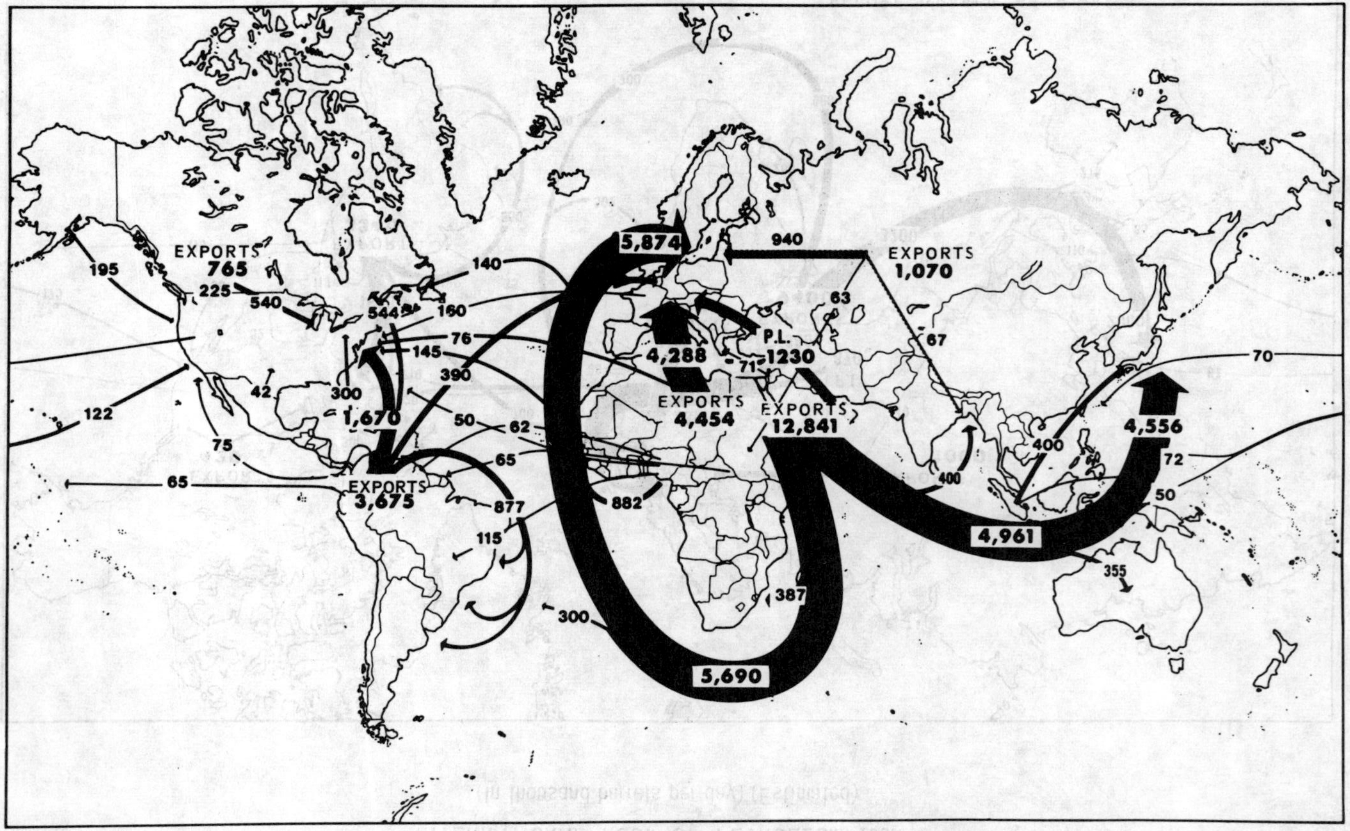

EXPORTS 1,070

EXPORTS 12,841

EXPORTS 4,454

EXPORTS 3,675

EXPORTS 765

EXPORTS 1,670

70
72
50
4,556
355
400
4,961
87
63
400
P.L.
1230
715
387
940
5,690
5,874
4,288
882
300
140
62
65
877
300
76
50
65
115
160
390
145
544
300
42
540
225
195
75
65
122
10

1.   Excluding imports into the USSR and other Eastern European Countries, and the People's Republic of China.

122

Figure 7
TANKER VOYAGE CHARTER RATES AND AVERAGE FREIGHT RATE ASSESMENT (AFRA)

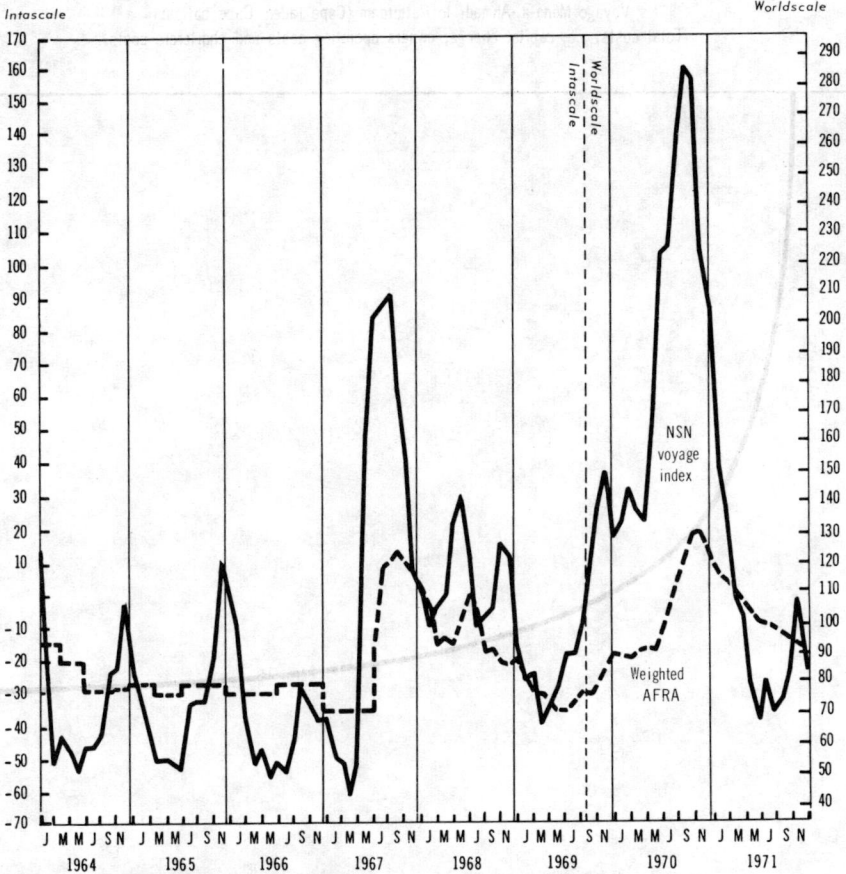

Note : Voyage Rates are derived from the Norwegian Shipping News monthly index.

Average Freight Rate Assessment is derived from the London Tanker Broker's Panel.

For a fuller description see note on page 142.

The assesments for the various size categories have been weighted together in proportion to their share in the world fleet, at the time of publication of the assessment

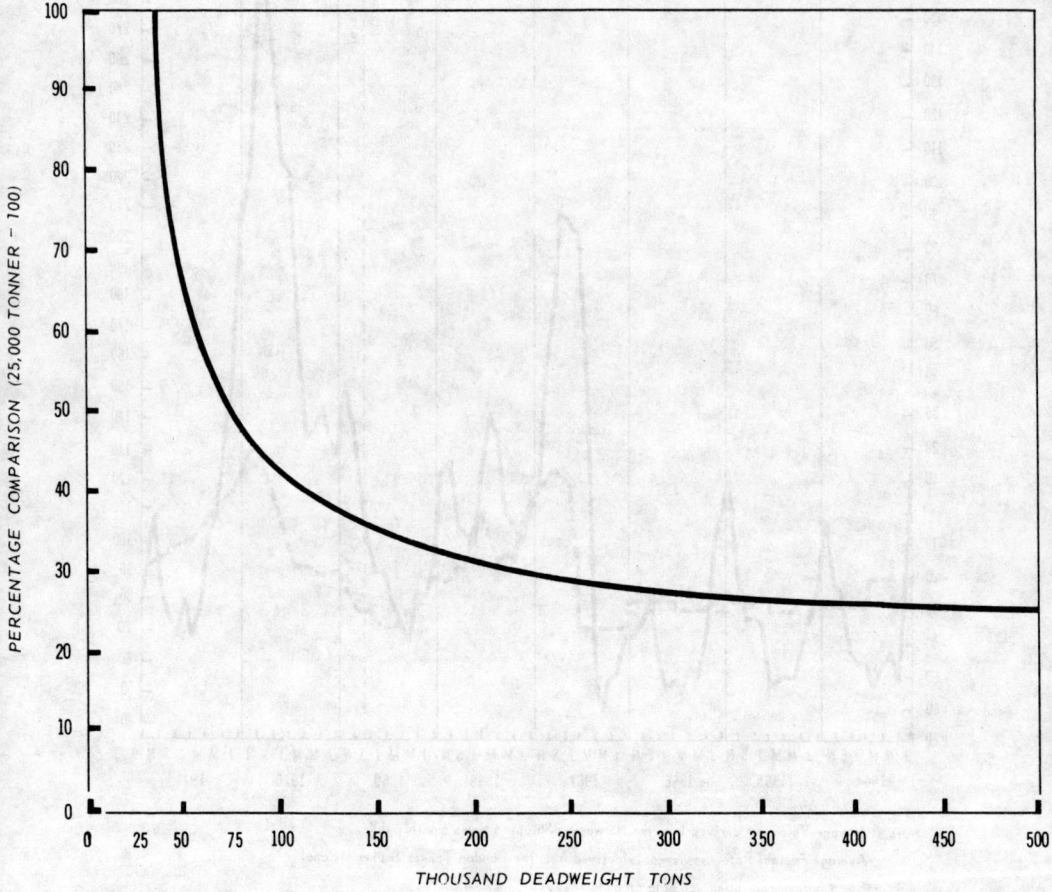

Figure 8

COMPARATIVE TRANSPORTATION COSTS BY VARIOUS SIZES OF TANKERS

Voyage Mena-al-Ahmadi to Rotterdam (Cape laden/Cape ballast)
Total costs, i.e. capital charge, owners operating costs and charterers costs

Source : Shell International.
May, 1970.

124

Figure 9

TRANSPORTATION COST FOR 16" DIAMETER PIPELINE

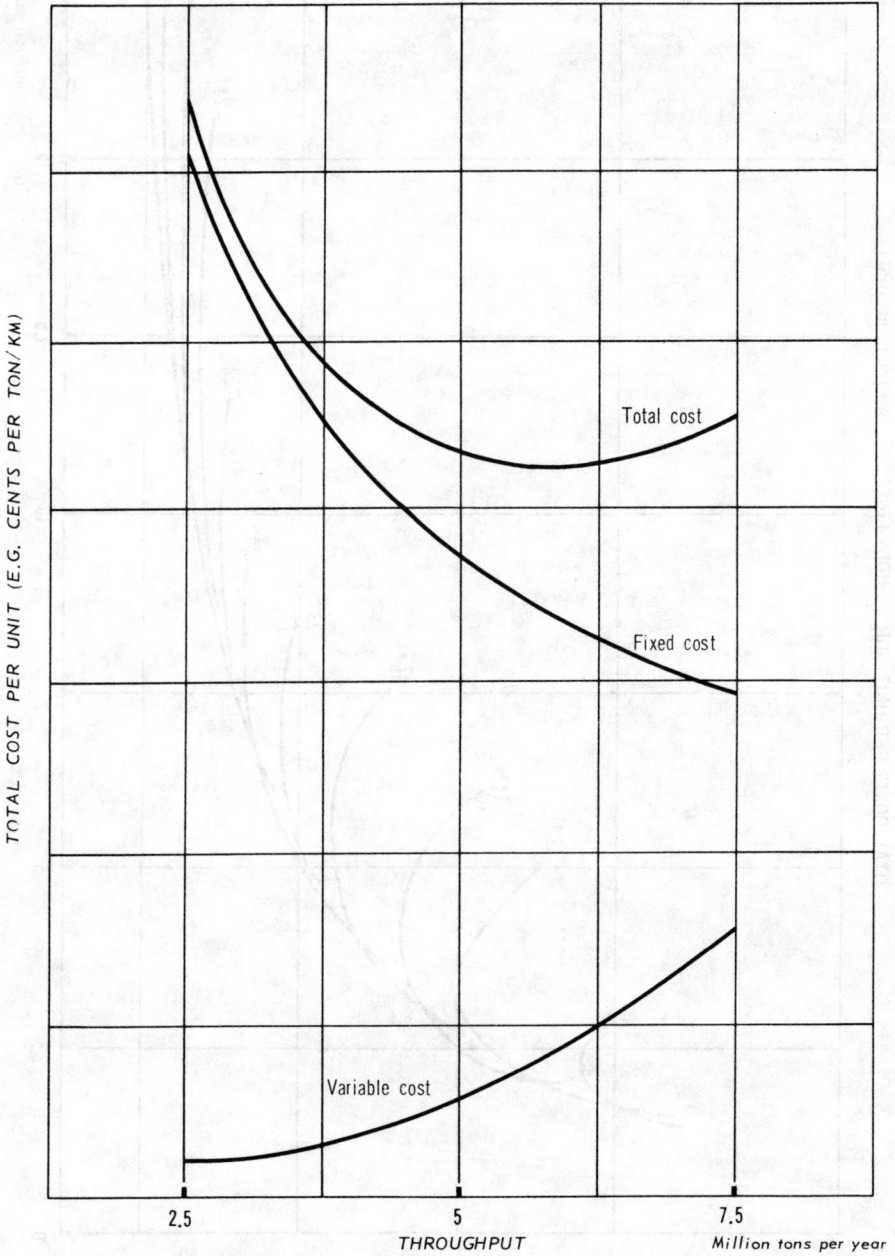

Total cost

Fixed cost

Variable cost

TOTAL COST PER UNIT (E.G. CENTS PER TON/KM)

2.5

5

7.5

THROUGHPUT

Million tons per year

Source : Shell Information Handbook, 1971-1972.

Figure 10

TYPICAL PIPELINE COST CURVES

TOTAL COSTS PER UNIT FOR VARIOUS LINE SIZES AND THROUGHPUTS

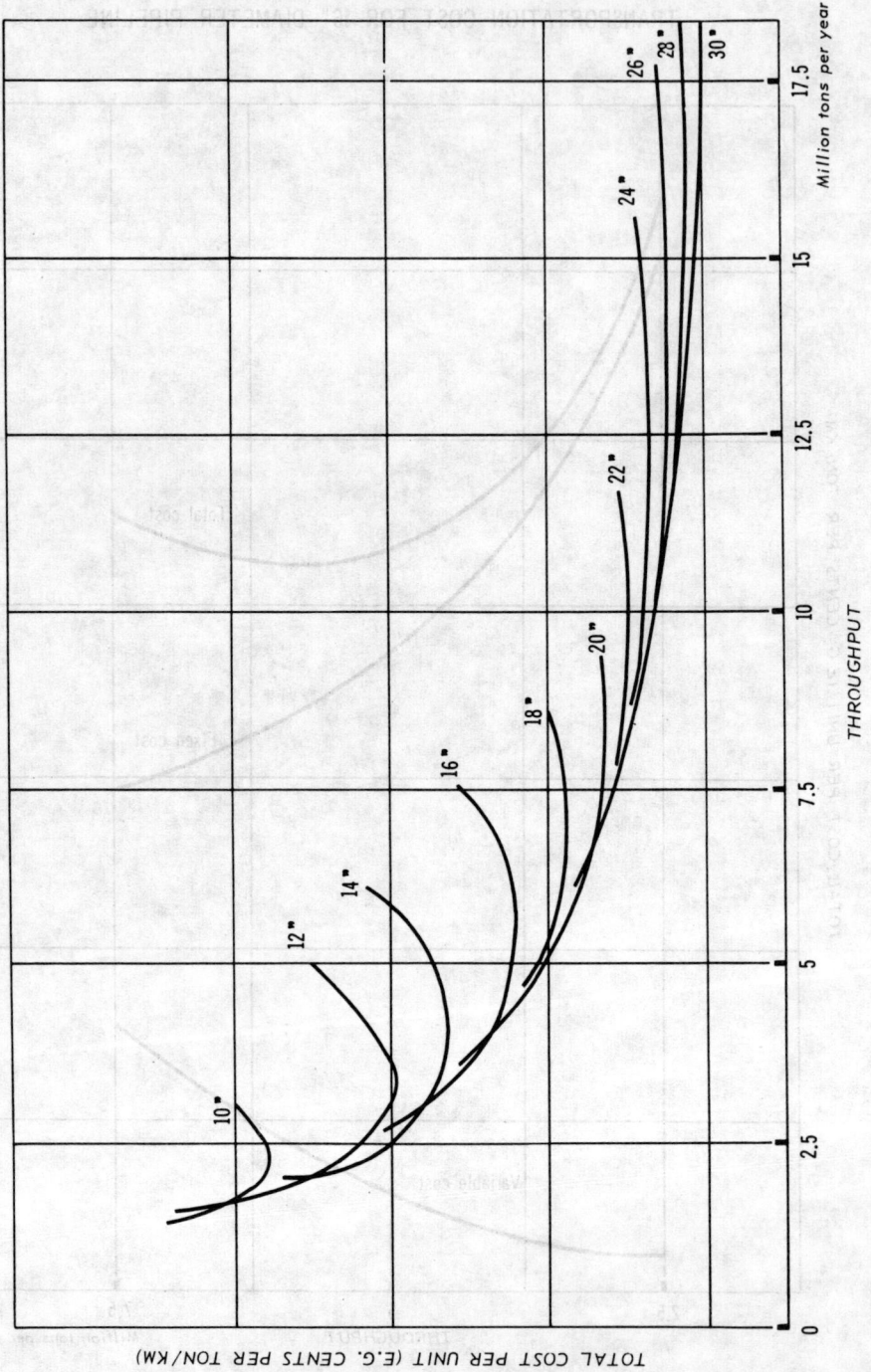

Million tons per year

THROUGHPUT

TOTAL COST PER UNIT (E.G. CENTS PER TON/KM)

10" 12" 14" 16" 18" 20" 22" 24" 26" 28" 30"

Source : Shell Information Handbook 1971-1972.

Table 25. WORLD TANKER FLEET AT END 1971[1]
(excluding combined carriers)
BY AGE, SIZE AND PROPULSION

Million deadweight tons

SIZE IN '000 d.w.t.	YEAR OF CONSTRUCTION							PROPULSION		NEW BUILDING IN PROGRESS AND ON ORDER AT END 1971[2]
	UP TO END 1945	1946-1950	1951-1955	1956-1960	1961-1965	1966-1971	TOTAL	MOTOR	OTHER	
Under 25	4.4	1.3	8.2	7.4	2.5	4.7	28.5	19.9	8.6	1.0
25- 45	1.0	1.3	5.0	14.7	4.1	1.9	28.0	8.1	19.9	3.4
45- 65	0.1	-	0.9	5.1	14.6	1.9	22.6	7.4	15.2	-
65-125	-	-	-	2.2	14.0	23.2	39.4	24.1	15.3	3.6
125-205	-	-	-	-	0.1	11.9	12.0	5.4	6.6	5.0
205-285	-	-	-	-	-	42.5	42.5	1.9	40.6	65.2
285 and over	-	-	-	-	-	2.3	2.3	-	2.3	10.4
Total	5.5	2.6	14.1	29.4	35.3	88.4	175.3	66.8	108.5	88.6

PROPULSION

Motor	0.8	1.1	7.1	8.9	17.6	31.3	66.8
Other	4.7	1.5	7.0	20.5	17.7	57.1	108.5

1. Excludes 20.3 million d.w.t. combined carriers,
2. Excludes 21.5 million d.w.t. combined carriers,
SOURCE: BP Statistical Review of the World Oil Industry - 1971.

Table 26.   WORLD TANKER FLEET, BY SIZE, FOR YEARS 1960-1971

Million deadweight tons

WORLD TANKER FLEET BY SIZE	1960	1961	1962	1963	1964	1965	1966	1967	1968	1969	1970	1971
Under 25,000 d.w.t.	36.8	35.6	34.0	32.7	31.3	31.2	30.0	29.7	29.8	28.9	28.3	28.5
25 - 45,000 d.w.t.	21.6	23.3	24.7	25.4	25.4	25.5	25.3	25.0	25.6	26.7	27.4	28.0
45 - 65,000 d.w.t.	4.3	6.0	8.4	12.0	16.1	19.3	21.2	21.8	22.1	22.0	22.5	22.6
65 - 85,000 d.w.t.	0.5	1.3	1.8	2.6	4.8	8.1	12.7	14.9	15.7	16.5	17.1	⎫ 39.4
85 - 105,000 d.w.t.	0.7	0.7	1.0	1.8	3.5	5.0	6.6	10.5	12.9	14.2	15.4	⎭
105 - 125,000 d.w.t.	0.1	0.2	0.3	0.4	0.4	0.9	2.5	3.5	4.8	5.6	6.0	12.0
125 - 205,000 d.w.t.	-	-	0.1	0.1	0.1	0.1	0.9	2.5	5.4	8.2	10.3	12.0
205 - 285,000 d.w.t.	-	-	-	-	-	-	0.2	-	2.6	11.1	26.7	⎫ 44.8
285,000 d.w.t. and over ...	-	-	-	-	-	-	-	-	0.6	2.0	2.0	⎭
TOTAL	64.0	67.1	70.3	75.0	81.6	90.1	99.4	107.9	119.5	135.2	155.7	175.3

SOURCE: B.P. Statistical Review of the World Oil Industry - 1971.

128

Table 27.   WORLDSCALE NOMINAL TARIFF

(as if 1st January, 1972)

	$ U.S. per Long Ton
To Rotterdam from:	
Ras Tanura, Saudi Arabia	9. 76
Kharg Island, Iran	10. 06
Ras Lanuf, Libya	3. 10
Bonny, Nigeria	4. 54
Punta Cardon, Venezuela	4. 28
To Yokohama from:	
Ras Tanura, Saudi Arabia	5. 77
Kharg Island, Iran	6. 07
Bonny, Nigeria	9. 97
Dumain, Indonesia	3. 15
To Philadelphia from:	
Ras Tanura, Saudi Arabia	10. 06
Kharg Island, Iran	10. 36
Ras Lanuf, Libya	4. 55
Bonny, Nigeria	5. 02
Punta Cardon, Venezuela	2. 04

Table 28. WORLD TANKER FLEET AT END 1971
(Excluding 20.3 million d.w.t. combined carriers)
(2,000 d.w.t. and over)

BY FLAG AND BY OWNERSHIP

FLAG	OWNERSHIP MILLION DEADWEIGHT TONS				TOTAL 1971	TOTAL 1970	CHANGE 1971 OVER 1970	SHARE OF TOTAL 1971
	OIL COMPANY	PRIVATE	GOVERNMENT	OTHER				
Liberia	8.2	34.7	-	0.2	43.1	37.8	+ 5.3	25
Norway	0.2	18.9	-	-	19.1	17.2	+ 1.9	11
United Kingdom	17.6	7.3	0.2	0.1	25.2	21.9	+ 3.3	14
Japan	3.1	15.7	-	-	18.8	15.6	+ 3.2	11
United States	4.0	4.0	1.7	-	9.7	9.5	+ 0.2	5
Panama	3.3	2.4	-	-	5.7	5.6	+ 0.1	3
France	4.7	2.8	0.1	-	7.6	5.8	+ 1.8	4
Greece	-	8.5	-	-	8.5	7.8	+ 0.7	5
Other Western Europe ..	10.5	13.2	0.1	-	23.8	21.4	+ 2.4	14
Other Western Hemisphere	2.7	0.3	0.3	-	3.3	3.4	- 0.1	2
USSR, Eastern Europe and China ...	-	-	6.2	-	6.2	6.1	+ 0.1	4
Other Eastern Hemisphere	1.0	3.2	0.1	-	4.3	3.6	+ 0.7	2
Total	55.3	111.0	8.7	0.3	175.3	155.7	+ 19.6	100
Fleet as at end 1970 ...	49.2	97.6	8.7	0.2	155.7			
Net increase 1971	6.1	13.4	-	0.1	19.6			

SOURCE: B.P. Statistical Review of the World Oil Industry 1971.

Existing Fleet

WORLD TANKER FLEET BY SIZE	COMPANY		PRIVATE OWNERS		TOTAL	
	No.	Dwt.	No.	Dwt.	No.	Dwt.
Under 25,000 dwt.	548	9,475,301	697	13,210,815	1,245	22,686,116
25- 45,000 dwt..	308	10,434,005	509	17,188,978	817	27,622,983
45- 60,000 dwt..	150	7,661,022	199	10,417,053	349	18,078,075
60- 80,000 dwt..	79	5,542,558	181	12,557,662	260	18,100,220
80-100,000 dwt..	42	3,791,977	131	11,841,544	174	15,633,521
100-125,000 dwt..	28	3,144,413	64	6,958,443	92	10,102,856
125-200,000 dwt..	20	3,317,982	50	7,676,497	70	10,994,479
200-300,000 dwt..	71	15,976,133	122	27,287,169	193	43,263,302
300,000 and over .	1	372,698	6	1,960,212	7	2,332,910
Total	1,247	59,716,089	1,960	109,098,373	3,207	168,814,462

New Buildings on Order

	COMPANY		PRIVATE OWNERS		TOTAL	
	No.	Dwt.	No.	Dwt.	No.	Dwt.
Under 25,000 dwt.	24	513,300	21	470,000	45	983,300
25- 45,000 dwt..	58	1,690,000	59	1,761,000	117	3,451,000
45- 60,000 dwt..	-	-	-	-	-	-
60- 80,000 dwt..	1	66,000	5	345,200	6	411,200
80-100,000 dwt..	2	196,599	5	432,300	7	628,800
100-125,000 dwt..	10	1,166,000	12	1,374,000	22	2,540,000
125-200,000 dwt..	9	1,221,600	27	3,746,600	36	4,968,200
200-300,000 dwt..	92	23,175,200	170	42,598,600	262	65,773,800
300,000 and over .	22	6,999,600	8	2,848,800	30	9,848,400
Total	218	35,028,200	307	53,576,500	525	88,604,700

SOURCE: John I. Jacobs Co., Ltd., World Tanker Fleet Review.

Table 30. SOME IMPORTANT PIPELINES

CRUDE OIL

PIPELINE	ROUTE	DIAMETER (In.)	LENGTH (Miles)	CAPACITY (Thous. b/d)	DATE OF COMMISSIONING
Interprovincial	Canada: Edmonton to Toronto and Buffalo (N.Y.)	16-48	3,553	1,300	1953
Comecon	USSR Hungary East Germany Poland Czechoslovakia	20-40	3,300	400	1964
Trans-Siberian	Tuymazy to Irkutsk	28	2,283	340	1964
	Saudi Arabia (Abqaiq) to Lebanon (Sidon)	30/31	1,068	500	1951
Trans-Mountain	Canada: Edmonton to Vancouver	24/30	718	400	1953
Capline	U.S.A.: St. James to Patoka	40	630	500	1968
IPC	Iraq (Kirkuk) to Syria (Banias)	12/32	556	1,100	1952
	Iraq (Kirkuk) to Lebanon (Tripoli)		531		1934
Tipline	Gulf of Aqaba-Ashkelon	42	160	460	1969
South European	France (Lavera) to FRG (Karlsruhe)	34/40	485	1,400	1963
TRAPSA	Algeria (Amenas) to Tunisia (La Skhirra)	24	482	240	1960
SOPEG	Algeria: Hassi Messaoud to Bougie	22/24	411	360	1959
	Libya: Sarir to Tobruk	34	317	600	
Transalpine	Italy (Trieste) to FRG (Ingolstadt)	40	290	700	1967
RRP	The Netherlands (Pernis) to FRG (Frankfurt)	24	285	700	1960
NWO	FRG: Wilhelmshaven to Cologne	28/40	223	900	1959
AWP	Austria: Wurmlach to Vienna	18	251	110	1970
RAPL	The Netherlands: Rotterdam to Antwerp	34	65	480	1971
	USSR: Aleksandrovskoye to Anzhero-Sudzhensk	48	500	1,000	1972

OIL PRODUCTS

PIPELINE	ROUTE	DIAMETER (In.)	LENGTH (Miles)	CAPACITY (Thous. b/d)	DATE OF COMMISSIONING
Colonial	USA: Houston to New York	30/36	3,110	1,200	1964
Plantation	USA: Lousiana to North Carolina	10/18	1,260	200	
RMR	The Netherlands (Rotterdam) to FRG (Frankfurt)	18/24	370	230	1966
UKOP	United Kingdom: Thames to Midlands and Mersey	6/14	245	200	1968
SPMR	France (Lavera, Dijon) to Switzerland (Geneva)	12/16	360	90	1969
TRAPIL	France: Le Havre to Paris	10/20	145	400	1953
	United Kingdom: Fawley to West London	10/12	64	140	1963
Explorer	USA: Houston to Chicago	12/28	1,300	263	1971

132

Table 30. (Contd.)
NATURAL GAS

PIPELINE	ROUTE	DIAMETER (in.)	LENGTH (miles)	CAPACITY (10⁶xft.3/d)	DATE OF FIRST COMMISSIONING
Trans-Canada	Canada: Burstal to Montreal	24-34	2,294	800	1959
Mid-America (LPG)	USA: Texas to Milwaukee/ Minneapolis	8/10	2,174	60,000 b/d	1960
	USSR: Central Asia to Moscow	40-48	1,700	2,200	1967
Little Inch	USA: Baytown to Bayway	20	1,475		1942
Pacific Gas	Canada (Alberta) to USA (California)	30-36	1,366	415	1962
	USSR: Bukhara to Urals	40	1,360	1,700	1963
Big Inch	USA: Longview to Bayway	24	1,254		1942
	Argentina: Campo Duran to Buenos Aires	22/24	1,083	243	1960
	Argentina: Comodoro Rivadavia	30	1,045	350	1964
	United Kingdom: West Sole to Easington offshore	16	45	200	1967
	USSR: Ukhta to Torzhok	48	800	1,450	1970
	USSR to Czechoslovakia (Bratislava) and Austria (Vienna)	28-20	300	400	1967
	Iran (Agha Jari) to USSR (Ostara)	42/40	700	1,050	1970

133

Table 31. HARBOURS IN OECD MEMBER COUNTRIES[1] ACCESSIBLE FOR FULLY LOADED
TANKERS OF 200,000 DEADWEIGHT TONS AND LARGER AND POSSIBLE EXPANSION PLANS

(Size in thousand dwt.)

	AVAILABLE	SIZE OF TANKER WHICH CAN BE ACCOMMODATED	EXPANSION PLANS	SIZE OF TANKER TO BE ACCOMMODATED
Canada	Come by chance	-	-	-
	Mispec Point	-	-	350
	Point Tupper	325	-	-
	Saint John	300	-	-
France	Dunkerque	-	1974	300
	Fos	300	1976	500
	Le Havre	250	1975	500
Germany	Wilhelmshaven	200	1973	250
Ireland	Bantry Bay	325	-	-
Italy	Augusta	250	-	-
	Falconara	-	1973	300
	Gaeta	-	1973	200
	Genoa	-	1972	500
	Milazzo	300	-	-
	Porto Torres	300	-	-
	Ravenna	-	1973	250
	Sarroch	200	-	-
	Taranto	300	-	-
Japan	Kawasaki	250	-	-
	Yokaichi	230	-	-
	Wakayama	200	-	-
	Chiba	200	-	-
	Tokuyama	210	-	-
	Nagoya	-	1973	250
	Oita	200	-	-
	Kagoshima	500	-	-
	Okinawa	325	-	-
	Tomakomai	-	1973	250
Netherlands	Rotterdam	250	1975	300
Norway	Slagentangen (Oslo)	250	-	-
Portugal	Sines	-	1976	350
Spain	Algesiras	200	-	200
	Bilbao	-	1975	350
	Cartagena	200	-	200
	Castellon	200	-	200
	Las Palmas	200	1974	350
	Malaga	-	1974	250
	Tarragona	-	1975	200
	Tenerife	200	1976	350
Sweden	Gotenborg	250	-	350
	Brofjorden	-	1974	200-500
Turkey	Izmit	250	-	-
United Kingdom	Clyde	-	-	320
	Finnart	300	-	-
	Humber River	210	-	-
	Liverpool Bay	-	-	300
	Liverpool Tranmere	-	-	200
	Maplin Sands	-	1976	500
	Milford Haven	250	-	-
	Tees Seal Sands	-	-	250
	Thames Haven	-	-	200-500
United States	Machiasport	-	-	300
	Virgin Islands	-	-	250
Brazil	Tramandai	200	-	-
Singapore	Singapore	250	-	-
South Africa	Durban	200	1973	300
Taiwan	Kaohsiung	250	-	-
Yugoslavia	Rijeka	-	-	350

1. And certain non-Member countries.

WORLDSCALE AND INTASCALE

Worldscale is the code name for the freight scale currently in use in the world tanker market, the objective of the scale being to provide a yardstick which accurately reflects the relationship between one voyage and another on all voyages on which tankers ply. It was preceded by a two-schedule system, of which Instascale was more widely used and had been employed since 1962.

The rates are set by assuming a notional ship (19,500 t. summer deadweight) with certain performance characteristics and a fixed charter cost per day as factors to calculate the cost of carrying a ton of oil on this vessel on a particular voyage. All basic rates are calculated on the same basis and a "Schedule" is built up. By this means the Worldscale schedule of freight rates provides a basic freight rate for every voyage which tankers are able to perform on a comparable basis. It supplies the mechanism for the working of the tanker market and a convenient framework for the evaluation of all aspects of freighting costs and charges.

AVERAGE FREIGHT RATE ASSESSMENT (AFRA)

AFRA is a method of assessment for charging freight which is being increasingly used by major oil companies. It became obvious in the early 1950s that to base freight charges on single voyage market rates was unfair, especially to large long-term customers. The London Tanker Brokers Panel provided an assessment, calculated according to specific terms of reference, of the average rate at which all vessels over 10,000 t. summer deadweight were operating on commercial charter during a given period. The terms of reference are revised from time to time when developments within the tanker industry necessitate changes. At present AFRA is published on the first of each month and the assessment relates to vessels on charter and in service during the month terminating on the fifteenth day of the previous month, e.g. AFRA published 1st October relates to vessels in service during the period mid-August to mid-September.

VII

ENVIRONMENTAL PROBLEMS

A. INTRODUCTION

309.   Industrial expansion, urban growth and increasing population and the accompanying air, land and water pollution and the degradation of our environment have increasingly become important social issues.   Not only does pollution intrude upon our esthetic values, but it as well has injurious effects upon human, animal and plant life and it also reduces the productivity of capital.   Man and nature share in contamination of the environment; man in his search for quality in life, nature by its very existence.   Industrial growth, upon which economic expansion has been based and which has been the foundation for improved quality in life, is directly dependent upon the utilization of increasingly greater amounts of energy.   For the first part of this century, coal was the primary form of energy; now oil and gas have become the dominant fuels.

310.   Pollution attributable to the operations of the petroleum industry is but a part of the total; the pollution generated by the use of petroleum products is equally small, but both are very apparent to the private and public sectors of society.   Pollution in its many forms is an ever-present potential at each stage of the oil industry, from the initial search for oil to the transport of petroleum products to the final consumer.   But recognition of the potential, and the implementation of safeguards at every step is a growing function of the oil industry.   At the final consumer, pollution derives from the combustion of fuels by stationary and non-stationary— e.g. automobiles— installations, as well as from the disposal of waste oils.   National and local governments and institutions increasingly are taking steps to regulate the use of fuels and to control the impact of this use on the environment.

B. THE PETROLEUM INDUSTRY

PRODUCTION

311.   During exploration (drilling) operations the main danger of polluting soil and water arises from the possible loss of drilling mud and the escape of oil.   Rigorous precautions, however, combined with careful geological surveys, can virtually eliminate this risk.   If a well passes through layers which are very sensitive for pollution by oil, such as water bearing

137

strata or very porous layers, it is cased and cemented to seal these layers against intrusion of oil. Both oil pollution and loss of production are so avoided. Once a well is completed and in operation, disposal of the so-called "connate" water that usually is associated with oil deposits becomes a major problem. The quantity of this water may form as much as 90 per cent of the liquid volume produced from some wells. A very efficient way of disposal on land is by re-injection into suitable subterranean formations, to avoid spoiling fresh water supplies. The water has to be treated to avoid complications from corrosion blocking the formation, bacterial growth and the like. Salt water obtained from demulsification and desalting processes can be dealt with in the same way.

REFINERIES

312.   The requirements relative to the maximum degree of contamination of refinery effluents by pollutants vary from location to location. For a few coastal refineries the only requirement is that there should be "no visible oil" in the effluent. Where effluent is discharged into an inland watervay or a lake, the effluent may affect the taste and odour of the water, which may make it less suitable for preparing drinking water and for fishing; requirements in such cases will be more severe.

313.   The most common contaminant in refinery effluents is oil. Other pollutants are suspended solids and dissolved components such as sulphides, phenolic— and nitrogen compounds and spent acid. The effluent itself partly exists of cooling water and partly of process water. Technically it is possible to remove all substances mentioned to a very high degree, by physical, chemical and biological means.

314.   Improvement in the efficiency of processing has been a very favourable factor in effluent amelioration. Probably the most important improvement has been the replacement of chemical treatment by hydrodesulphurization for removal of sulphur components from distillates. Another major improvement in new refineries has been a proper lay-out and well-designed drainage system.

315.   Once-through cooling water systems are most commonly used in coastal refineries. There always exists the possibility of contamination of the cooling water by oil due to leakages, which are very difficult to prevent. In locations, mostly inland, where traces of oil in the effluent are not tolerated or where the supply of water is limited, the amounts of cooling water can be considerably reduced by the use of circulating cooling water systems or air cooling. Air cooling has added to the noise level, but steps are being taken to control it. Also many methods for reducing the use of process water are being applied. At such refineries oil separators are much more effective and chemical and biological treatment can be applied.

316.   A number of procedures for treatment of the effluent are available:
    — stripping with flue gases or with fuel gas and stream, for the removal of sulphides, lower mercaptans and ammonium;
    — gravity seperation to remove most of the oil and suspended solids;

— chemical flocculation or oil flotation for further removal of oil and suspended solids;

— biological purification to remove oxygen-consuming contaminants, including phenolic material.

Handling and disposal of sludge obtained from chemical and biological treatment are difficult. Sometimes the dewatered sludge is used as land-fill, or a sludge may be thickened and incinerated.

Performance of refineries with respect to water pollution

317.   Concentrations of contaminants are the most common properties measured. It is generally realized, however, that the total amount of load of contaminants in effluents is more important than concentration. This is because the concentrations in refinery effluents seldom reach a level where the effluent has an acute toxicity for aquatic life. It is the total amount of contaminants that is important. Table 32 surveys what has been achieved so far in Europe. This survey covers about 80 per cent of refinery throughput in Western Europe[1]

As can be seen, major improvement in reducing the load of contaminants in refinery effluents has been secured at those refineries constructed after 1960.

318.   Data on waste loads at US refineries have been published by the American Petroleum Institute[2], and comparison between the US and Europe is given in Table 33. The data are classified according to the method of effluent treatment. Category III is used in most European inland refineries.

319.   With respect to thermal pollution, it may be mentioned that biologists and ecologists consider it necessary that to support normal aquatic life in surface water, the temperature should remain below 25-30°C. The temperature of refinery effluents is nearly always below 30°C and the temperature rise of the water after mixing of the effluents and surface water in essence can be ignored.

320.   A recent American Petroleum Institute survey indicates that the expenditures for air and water conservation by the US petroleum industry have more than doubled since 1966[3]. Total expenditures for air and water conservation by the US petroleum industry (in million of dollars):

	Air	Water	Total
1966	126	146	272
1967	167	191	358
1968	187	205	393
1969	231	224	455
1970 (est.)	271	288	559

1. P.C. Blokker and M.J. Marcinowski, Survey on quality of refinery effluents in Western Europe, Petrol. Review 25 (1971), 39 5/399.

2. Crossley S.D. Surveys Inc. Report of the API Committee for air and water conservation, 1967; Domestic refinery effluent profile, Sept. 1968.

3 "Report on Air and Water Conservation Expenditures of the Petroleum Industry of the United States, 1966-70", API Publication No 4075, (February 1971).

During these years more than half the capital expenditures were directed to controlling sulfur oxides.

321.   The major potential refinery emissions which may contribute to air pollution are sulfur compounds, hydrocarbons, nitrogen oxides, particulates— including smoke, and carbon monoxide.   The character and quantity of refinery atmospheric emissions vary greatly from refinery to refinery, and reflect among other things the age of the refinery, the types of crude oils being processed, those air pollution control measures being used at the refinery, the complexity of the processes employed and the general state of maintenance and housekeeping procedures being followed.

Amounts of air pollutants emitted from refineries

322.   In the US those refineries which were not subject to any regulations produced emissions of approximately 21 t of hydrocarbons, 46 t of carbon monoxide and 3 t of particulates per day per one million tons of crude oil processed per annum; in refineries subject to rigorous regulations the respective emissions are 1.5 t of hydrocarbons, 2.7 t of carbon monoxide and less than 0.3 t of particulates (1968).   Comparable data are not available for European refineries although studies are now being made, especially in Germany.   Sulphur oxides and nitrogen oxides originate at refineries from the combustion of oil and gas in heaters and boilers and also from the refinery processes itself.   The amounts of nitrogen oxides are relatively small, but the emissions of sulphur oxides may be substantial, depending on the sulphur content of crude oil used at the refinery.   A typical refinery burns 7 per cent of its throughput for the heating purposes, approximately 70 per cent of which will be in the form of gas.   The remaining 30 per cent will be heavy residual oil, which is usually high in sulphur content.   The emission of sulphur oxides and nitrogen oxides from refining has been examined in some detail by the Joint Ad Hoc Study by the OECD Environment Committee, Energy Committee and Oil Committee on Air Pollution from Fuel Combustion in Stationary Sources.

Hydrocarbons and malodorous gases

323.   Pure hydrocarbons have a low toxicity and little smell.   However, some unsaturated hydrocarbons may cause, together with other pollutants under the influence of solar irradiation, the formation of aerosols, thus giving rise to a faint haze (Los Angeles smog).   Moreover, hydrocarbon vapours often contain foul-smelling compounds.   Rigorous control is therefore applied to hydrocarbon vapour losses.   Gasoline and crude oil are nowadays almost always stored in floating roof tanks, which reduces the emission of volatile hydrocarbons and any associated odorous compounds by 85 per cent or more.

324.   In modern refineries considerable improvement in handling malodorous gases has been achieved with removal of sulphur compounds from distillates by hydrogen treatment.   The resulting gas streams containing a

high concentration of hydrogen sulphide are directed to a sulphuric acid plant or a sulphur recovery plant.

325.   Various processing operations, especially catalytic cracking produce waste water solutions, generally called "sour waters". These waters often have a repugnant smell. The usual method for removal of the odorous compounds is to strip the water with flue gases or steam and to burn the off-gases or lead them to a sulphur recovery unit.

326.   Open refinery sewers, open oil separators, catalyst regenerators, leakages through the seals and glands of compressors, pumps and valves are other potential sources of offensive smell. There are various measures to resolve these problems, such as the use of covered sewers and oil separators, stripping the odorous compounds from the water before discharge into sewers, use of mechanical seals, and so forth. However, constant attention to careful operation and maintenance can in itself contribute substantially to the odor-free operation of a refinery.

TRANSPORT

327.   In transport, distinction can be made with regard to the possible risks of oil pollution between pipeline transport and transport by road or railway.

328.   In Western Europe, a pipeline system of about 12,500 kilometers is in operation in the civil sector and this system transported an average of about 220 million tons of crude oil and oil products per year during the period from 1966 up to 1971. This system has been evolved largely within the last decade. Generally, codes of practice for the construction and operation of a pipeline are established by or under the auspices of public authorities. These codes contain special requirements for safe operation (such as, for example, the obligatory appliance of cathodic protection against corrosion). Just 34 incidents were reported by the pipelines operated by oil companies in Western Europe from 1966 up to 1971. Nineteen of these were caused by accidental factors. The total oil spillage was approximately 2,400 cubic metres (including one major case in 1966 involving a spillage of 1,300 cubic metres) out of a total transported volume of about 1.2 billion cubic metres of crude oil and oil products (i.e. 2 parts per million). This low oil spillage demonstrates that a pipeline system is a very safe way to transport oil.

329.   Considering the extent of the US petroleum pipeline system and the volumes of crude and products moved, losses from spillage are remarkably low: approximately 6/thousandths of 1 per cent of the volume moved annually[1]. Principal causes of pipe-line accidents are dominated by external corrosion, with outside sources such as earth-moving equipment and operating error of much lesser importance. Natural catastrophies such as landslides, earthquakes and floods have been of minor magnitude in their effects on pipelines.

---

1. "Environmental Conservation, the Oil and Gas Industries", a report by the National Petroleum Council, Washington, DC, 1971.

330.   Today the growing importance of oil produced offshore has made the use of long and large submarine pipelines a common practice. Pipelines are usually laid under the sea bottom in trenches to protect them from currents and other external forces.

331.   When quantities of oil products to be transported and distances to be covered are relatively small, it is not feasible to use pipelines. In such cases transport must be realized by road, railway or inland waterway.

332.   Generally speaking, the high safety level of pipelines cannot be matched by those means of transport. However, statistics on the number of accidents, the amounts of specific products spilled and the causes are rarely available, and when they are, the information often lacks common data and is therefore difficult to compare. German authorities report that in 1967 in Germany about 1,400 t of oily liquids were spilled during transport of approximately 73 million tons of oil products, i.e. circa 20 parts per million. The majority of the oil spillage involves heating fuel, reflecting its wide distribution to a great number of small consumers, with delivery sometimes under arduous conditions.

333.   Loading and unloading operations are closely connected to transport incidents. The overwhelming part of these casualties is caused by human failure. The technical causes can be reduced considerably by special equipment like overfilling control devices, which are obligatory in certain countries. The oil companies have organized special courses to educate drivers involved in the transportation of petroleum products. It is also becoming normal for tank trucks to carry absorbing material for use in removal of small spills that may occur during delivery. Improvement in the safety of road transportation can also be expected by the adaptation of national legislation to the "European Agreement on Carriage of Dangerous Goods by Roads."[1]

Transport on inland waterways

334.   Transport on inland waterways mostly concerns oil products. Pollution can be caused by spillage or accidents; however, little information is available on this subject. Furthermore, inland waterways can be polluted indirectly, by accidents or spillage from elsewhere.

Transport at sea

335.   The major part of the pollution of the sea by oil is caused during transport activities. A small part originates from oil contaminated refinery effluents witch are discharged by coastal refineries. Figures of the latter are given in the relevant section.

336.   Over 600 million tons of crude oil were transported by sea to Western Europe in 1970, and more than 160 million tons to Japan. The

---

1. Working Party No. 15 of the International Transport Committee of the Economic Commission of Europe: European Agreement on Carriage of Dangerous Goods by Road (Annex B: Provisions concerning transport equipment and transport operation).

major cause of oil pollution of the sea results from the practice of discharging tank washings into the open water. Both the United Nations, through the Intergovernmental Maritime Consultative Organization (IMCO), and the private oil industry are taking steps to minimize oily discharges. The latter has developed the so-called Load On Top (LOT) system, whereby residues are retained on board during the process of tank washing. Only after the greater part of the oil has settled out is the water discharged, while the oil layer remains in the vessel. IMCO, in 1954, drafted an International Convention for the Prevention of Pollution of the Sea by Oil. This Convention was amended in 1962 and again in 1969. The current position for tanker operation is based on the 1954 Convention as amended in 1962, the limit of 0,01 per cent within designated zones still applying, but when the 1969 amendments are ratified and implemented the standards for tankers will be:

a) no oil discharge whatever will be permitted within certain specified zones (normally 50 miles from coastlines) ;

b) where discharge is permitted it may not be more than 60 litres per mile;

c) total discharge on any ballast voyage will be limited to a maximum of 1/15,000 of cargo capacity.

The 0.01 per cent criteria, with certain provisos, will still apply to ships other than tankers. The world-wide application of these amendments will reduce intentional operational discharge to a non-polluting level. It will clearly take some time to bring them into force; at present only seven of the required twenty-nine ratifications have been obtained. Major sectors of industry are however already applying and enforcing the 1969 amendments on their crude-carrying ocean tonnage, and about 90 per cent of refineries receiving crude from overseas are equipped to receive cargoes with LOT residues.

337.   Apart from discharge of tanker washings at sea, pollution can be caused by accidents. When loaded tankers are involved, pollution can be spectacular indeed. As regards such major oil spills, three areas of concern emerge: firstly, what can be done to prevent accidents at sea; secondly, once an accident has happened what can be done to reduce the damage to the beaches, shores and the sea-life; and thirdly, who can be held liable for the resultant damage. Much has been done to meet these concerns, again by IMCO as well as by the oil industry.

338.   The growing size of tankers has a favourable influence on the intensity of traffic at sea, as the number of vessels that are needed to transport a certain volume of oil is thus reduced. The chance of collision is therefore reduced. If, however, an accident should occur, the risk of an eventual massive oil spill is increased. To that effect, recently an agreement was reached within IMCO on a safer tanker design, having in mind a limitation of the tank size, of wing tanks and as well as of centre tanks, which would minimize oil spill in the event of accident. Also steps have been taken to delineate recommended routes in shipping lanes where density of tanker traffic is high.

339.   With regard to minimizing the possible damage on shore, IMCO has also framed a Convention which gives the right to a coastal state to

143

intervene on the high seas in cases of oil pollution casualties. The Convention authorises the state, whose coast is threatened by oil, originating from a ship that is involved in an accident to take appropriate measures against that ship. On a regional basis, North Sea states have entered into co-operation in tracing of and dealing with oil spills on the North Sea. Concerning the technical means of abatement of (sea-) water pollution, continuous research has been done by a great variety of industries and individuals. Nowadays, if circumstances are not too unfavourable, rather efficacious measures are available for clean-up of oil spills in open water, with only minor harm to sea-life.

340.   Liability for oil pollution damage by sea-going vessels is the subject of two other Conventions, again drafted by IMCO. In 1969 the Convention on Civil Liability for Oil Pollution was drafted and in 1971 a Convention on the Establishment of an International Compensation Fund for Oil Pollution Damage. The latter Convention can only be brought into force by states which have ratified the liability Convention of 1969. Both Conventions together, which are yet to be ratified[1], result in a combined, strict liability for the shipping industry (along the lines of the Convention of 1969) and the oil industry (according to the 1971 Convention), up to a limit per incident of $30 million, which amount can be doubled, if necessary, by those parties to the Convention of 1971. The strict character of the liability, the high monetary liability, the compulsory insurance scheme in the Convention of 1969 as well as the scheme of setting up an international Fund to which national oil industries in states parties to the 1971 Convention are obliged to contribute are all rather new and unusual elements in marine law and clearly demonstrate the interest of countries, working together in IMCO, in creating a just and workable scheme for compensation of victims of oil pollution at sea. However, it is likely that the entry into force of the Conventions will take some time.

341.   While IMCO was engaged in the above-mentioned Conventions, tanker owners and oil industry have voluntarily set up comparable schemes, namely the "Tanker Owners Voluntary Agreement Regarding Liability for Oil Pollution" (TOVALOP), which is subscribed to by 96 per cent of the non-Communist tanker tonnage, and a "Contract Regarding an Interim Sup-

---

1. The 1969 Convention on civil liability for oil pollution damage shall enter into force on the ninetieth day following the date on which governments of eight states including five states each with not less than 1,000,000 gross tons of tanker tonnage have either signed it without reservation as to ratification, acceptance or approval, or have deposited instruments of ratification, acceptance, approval or accession with the Secretary-General of IMCO.
The following governments are signators: Australia, Belgium, Brazil, Cameroon, Republic of China, Dominican Republic, Finland, France, Federal Republic of Germany, Ghana, Guatemala, Iceland, Indonesia, Ireland, Italy, Ivory Coast, Malagasy Republic, Monaco, Netherlands, Panama, Poland, Roumania, Spain, Sweden, Switzerland, UK, USA, Yugoslavia. All signators were subject to ratification or acceptance. The only contracting government is Senegal which acceded on March 27, 1972.
The international Convention on the establishment of an international compensation fund for oil pollution damage, among many other requirements, cannot enter into force before the liability convention has entered into force. To date, only the original signator governments as follows have signed the convention: Algeria, Belgium, Brazil, Federal Republic of Germany, Ghana, Poland, Portugal, Sweden, Switzerland, UK, USA and Yugoslavia. None have ratified the convention and there are no additional signators.

plement to Tanker Liability for Oil Pollution" (CRISTAL). Both schemes are already in operation and can therefore fruitfully overlap the period which is needed to bring the IMCO Conventions into force.

STORAGE

342. The widely scattered use of petroleum products for many purposes and by a varied range of consumers results in a steadily growing number of storage installations. Storage installations roughly can be distinguished between permanently supervised and/or operated installations and intermittently operated ones. Generally it can be said that the former represent a smaller risk than the latter. This explains the often-observed higher incident rate at domestic fuel installations compared with those in industry and commerce.

343. There is a wide variety of special measures to reduce the damage to ground and surface water: overfilling and leak detection devices, the bunds around the tanks and the double wall containers. However, normal technical improvement of the installations as such, for example the use of more resistant construction material, e.g. reinforced plastic, periodic cleaning and inspection, also contribute to the prevention of pollution and pollution damage. Often an impervious cover of special concrete paving is applied at all spots where oil can be spilled, in storage tank areas as well as in refinery process areas. Also, drain systems are used to collect all oil-contaminated water.

344. The very strong restrictions in some countries with respect to the location, construction and inspection of storage tanks lead one to expect that the relative number of incidents will decline, especially when the more or less unsafe older installations are gradually eliminated.

C. AIR POLLUTION BY STATIONARY INSTALLATIONS

345. Combustion of fuel-oil in stationary installations causes the emission of sulphur dioxide, nitrogen oxides and particulate matter, including soot. In the following, only a general out-line of the problems is presented. Additional, detailed information can be found in a special report on this subject[1].

SULPHUR DIOXIDE

346. The amount of SO_2 emitted can be calculated directly from the sulphur content of the fuel. With 2 per cent sulphur in the fuel the amount of SO_2 emission is 40 kg/ton of fuel consumed. Residual fuels may contain 0.5 per cent by weight up to 3.5 per cent, depending on the crude oil from which they originate. Lighter fuels generally contain less sulphur than heavy fuels. Sulphur content in the lighter fuels ranges from 0.3 per cent to 1.5 per cent by weight.

---

1. "Air Pollution from Fuel Combustion in Stationary Sources", OECD report.

347.   Consumption of heavy fuel-oil is largely restricted to refineries, electric power plants and major industries. Domestic consumers generally use the lighter fuels, but their waste gases are normally emitted at a much lower level than by industrial consumers and therefore may make a significant contribution to pollution levels. A survey of the quantities of fuel used in OECD countries in 1969, is given in Table 34[1].

348.   Reducing sulphur dioxide ground level concentrations is being achieved by using high stacks, by burning low sulphur fuels, or by desulphurization of stack gases.

349.   The use of high stacks may solve a local problem, but if the emissions are very large and meteorological conditions are very unfavourable the use of high stacks may occasionally not be sufficient.

350.   The lighter fuel-oil fractions have, as already noted, an very low sulphur content, as reduction during the refining process is now a standard practice. The sulphur content of residual fuel on the contrary is today dependent on the sulphur content of the crude oil from which it originates. If, as a consequence of more rigorous requirements on the emission of sulphur dioxide, the demand for low sulphur fuels considerably increases, there will not be sufficient crude oil amenable for producing low sulphur residual fuel. In that case, reduction of the sulphur content of these fuels must be accomplished. Although in principle technically possible, the means to do so is still in development and will be very expensive; one facility is in operation at Lake Charles, Louisiana, using the H-oil process. The reported costs for direct residual fuel-oil desulphurization are extremely variable and there are still many factors which are not adequately identified. Reported operating costs vary from $150 to $480 and capital costs from $380 to even $2,500 per ton of sulphur removed.

351.   Removal of sulphur dioxide from the stack gases is another potential route which may prove suitable for large installations. Again, reported costs for flue gas desulphurization fluctuate widely, depending on the nature of the process and the size of the unit, as well as on the unknowns involved in projecting from pilot to full scale operation and the use of variable economic factors. Reported operating costs vary from $35 to $290 per ton sulphur removed, and capital costs from $160 to $1,100. Most estimates are applicable only to very large units. For smaller units the costs increase rapidly with decreasing size so that it is quite likely that the fuel-oil desulphurization process will become the preferred route in most cases.

Nitrogen oxides (NOx)

352.   To some extent, the type of fuel burned has an impact on the emission of NOx, reflecting the percentage of nitrogenous organic compounds in the fuel. It should, however, be kept in mind that the greater part of the NOx emission originates from combustion of nitrogen from the air so that NOx emission is not specific for the combustion of oil products. It has been established that the concentration of NOx in the waste gases is very

---

1.   "OECD Statistics of Energy 1955-1969", Paris, 1971.

dependent on combustion conditions. These conditions are defined by such factors as flame temperature, amount of excess air, residence time of the combustion mixture, and so forth. A major survey on this aspect, conducted in the US[1] showed a clear difference between tangentially and horizontally fixed units, averaging concentrations of 200 ppm and 500 ppm NOx in the waste gases respectively. The ranges, however, from which these averages were drawn are so great that further study on this subject seems necessary.

Particulates, including soot

353.   The soot content of combustion gases from residual fuel is very dependent on the installation design, combustion conditions and the state of maintenance. For well-maintained and operated large units the soots content is low. The fly ash content is directly dependent on the ash content of the fuel, which is mostly below 0.05 per cent. This leads to an ash content of the combustion gases below 50 mg/Nm³ which is much lower than with coalfiring.

D. AIR POLLUTION BY NON-STATIONARY INSTALLATIONS

GASOLINE ENGINES

354.   Complete oxidation of gasoline yields only carbon dioxide and water but combustion of gasoline in an internal combustion engine also forms other products. These include carbon monoxide, unburned or partially burned hydrocarbons, oxides of nitrogen and particulate material consisting predominantly of lead compounds. Approximately 40 per cent of the hydrocarbon emissions from older vehicles derived from evaporative losses from the fuel system and blow-by gases emitted from the engine crankcase. Newer vehicles are equipped with positive crankcase ventilation systems, which eliminates about half the above emissions. The remaining emissions come from the exhaust pipe.

355.   The rate of emissions of these pollutants varies with operating conditions, i.e. idling, acceleration, deceleration or steady running. It is generally believed, however, that they contribute considerably to pollution as a whole and that steps to reduce these emissions are in order.

356.   It has been mentioned in this chapter that some hydrocarbons, together with nitrogen oxides, contribute to the formation of photochemical smog. Carbon monoxide in itself is very toxic, dependent of course on conditions of concentration and exposure. There is concern that undesired levels of atmospheric lead in congested urban areas may in the future reach hazardous levels during unfavourable atmospheric conditions. A greater problem with regard to lead, however, is that it is deleterious to the activity of certain catalysts, the use of which may be necessary in the future to achieve stringent levels of control over the other pollutants.

1. US Dept. of Health, Education and Welfare. Atmospheric Emissions from Fuel Oil Combustion 1968. Publ. No. 99-AP-2.

357.   The most rigid legislation on allowable emissions are or will be applied in the US.   In 1975-76 control levels will be imposed which will lead to about 98 per cent reduction of hydrocarbons and carbon monoxide and 90 per cent reduction of nitrogen oxides compared with uncontrolled emissions in the early 1960s.   Unleaded gasoline will be necessary to facilitate the use of sophisticated control techniques to reach these extreme requirements.

358.   The principal legislation in Europe is that developed by the UN Economic Commission for Europe, in Geneva, and subsequently adopted by the European Economic Commission in Brussels (Common Market). This limits the emission of carbon monoxide to less than 4.5 per cent by volume during idle operation, whilst crankcase emissions are restricted to 0.15 per cent of the total weight of fuel consumed.   The maximum permissible emissions of carbon monoxide and hydrocarbons from the tail pipe during the European Test Cycle are related to vehicle weight and they represent about a 20 per cent reduction on emissions from uncontrolled vehicles. There are indications that these limits, which apply only to new vehicles, will be reduced by an additional 20-25 per cent by 1974.   The techniques to achieve these lower levels of control in Europe are unlikely to require any significant changes in gasoline, including the lead alkyl content.   No limits have been set for nitrogen oxides or particulate emissions in Europe.

359.   Some restrictions are being placed on lead in Europe, particularly in Western Germany where a limit has been set of 0.4 gram of lead per litre for 1972 and 0.15 gram for 1976.   This reflects efforts, in the interest of public health, to avoid a rise of noxious lead emissions by motor vehicle engines in connexion with the increasing number of motor vehicles, after all known methods had proved unreliable or uneconomic.   At the same time this is a pre-condition for the application of certain catalytic afterburner systems which can be operated only with gasoline with low-lead content as a possibility for a far-reaching purification of exhaust gases in the case of motor vehicle engines.   The provision issued may, however, lead to an increase in manufacturing costs, changes in gasoline composition as well as to a slight reduction in the Research Octane Number and/or the Motor Octane Number.   Fears that as a consequence existing vehicles might no longer be operated satisfactorily have, however, proved unfounded as far as the first stage is concerned.   At present the petroleum industry is considering economically justifiable possibilities for the realization of the second stage.   If, indeed, the removal of lead would be achieved only by allowing octane levels to drop, and adjusting vehicle engines accordingly, the increase in consumption might be 8 to 12 per cent— equivalent, for OECD Europe in 1976, to an additional requirement in the region of 10 million tons of gasoline.

360.   Preliminary estimates of increased investment costs to produce gasolines satisfying the West German lead specification have been made. One study has been made on the basis of supplying an estimated 60 million tons of gasoline in 1976 to the six countries of the Common Market.   The production, with a lead content of 0.15 gram per litre, of 98-octane Premium grade gasolines and 92-octane Regular gasoline for the provision of the six countries would require an investment of $470 million.   For a zero-lead content this investment would rise to $1,560 million (1970 US dollar

values with some allowance made for inflation). The above estimates are based on the use of processing routes commonly in use in European refineries for gasoline production, i.e. catalytic reforming, leading to a certain increase in the aromatic content. In the event that the industry might be obliged to use more unusual processes to produce gasoline, such as alkylation and isomerization, to meet requirements relating to aromatic content of gasolines, then the investment costs would be very much higher.

Diesel engines

361.   Because of the great amount of excess air, fuel combustion in a Diesel engine is much more complete than in the gasoline engine. On the average, the amounts of carbon monoxide and hydrocarbons, relative to the fuel supplied, are approximately ten times less than those of present gasoline engines (comparison of concentrations is misleading because of the high dilution of Diesel exhaust by the excess air), but NOx emissions are greater than those of gasoline engines.

362.   The principal nuisances arising from the exhaust of diesel engines are, however, black smoke and odour. Black smoke may develop when the maintenance of the engine is insufficient but also when too much fuel is delivered to the engine. This can be controlled by proper maintenance and restricting the power extracted from the engine, and to a certain extent by fuel additives. Visible smoke is always accompanied by offensive odour but a virtually clear exhaust nevertheless may carry offensive odours. Attempts to associate factors in engine design or fuel composition with the odour problem up till now have failed.

363.   A regulation limiting the opacity of diesel exhausts has been developed by the Economic Commission for Europe and is under consideration by the Common Market.

Waste oil

364.   Rather serious pollution can be caused by the uncontrolled discharge of waste oil products from garages, service stations, private consumers, industries and others. In some countries, schemes and organizations are set up to collect these waste oils for controlled combustion or even, if technically and economically feasible, for regeneration.

365.   The foregoing has shown that much attention is being given to the problems of pollution related to handling and use of oil and oil products. Nevertheless, especially in the matter of air pollution, much still can be done. In most cases, the technical means for reducing pollution to acceptable standards are already available or will become so within a relative short term. It must be stressed, however, that application of these means will require very great investment costs. It seems to be preferable, therefore, that legal measures from the side of governments and the technical and economical possibilities for the abatement of pollution be kept in harmony as much as possible.

Table 32. WEIGHTED AVERAGE LOADS* FOR MAIN CONTAMINANTS IN REFINERY EFFLUENTS
IN WESTERN EUROPE

| LOCATION | PERIOD OF CONSTRUCTION | THROUGHPUT (MILLION TON/YEAR) | Contaminants in kg/1,000 tons/Crude Oil | | |
			ORGANIC MATERIAL "BIOLOGICAL OXYGEN DEMAND"	OIL	PHENOLS
Coastal	Prior to or in 1960	212	420	155	2.2
	After 1960	66	41	62	1.3
Inland	Prior to or in 1960	54	375	361	1.8
	After 1960	61	51	11	0.32

* Sum of crude throughputs x load per 1,000 ton crude.

150

Table 33. WEIGHTED AVERAGE OF WASTE LOADS IN EUROPE AND THE UNITED STATES, AS A FUNCTION OF EFFLUENT TREATMENT

Kilograms per 1,000 tons of Crude Oil Intake

EFFLUENT TREATMENT	BIOLOGICAL OXYGEN DEMAND		OIL		PHENOL	
	US	EUROPE	US	EUROPE	US	EUROPE
I. Gravity separation only	350 (178–485)	470 (113–503)	178 (82–221)	189 (26–268)	29 (3–43)	2.1 (0.9–2.4)
II. Gravity separation + chemical treatment	165 (89–294)	180 –	46 (3–56)	27 –	15 (3–20)	2.3 –
III. Gravity separation + chemical treatment + biological treatment	172 (86–357)	97 (4–150)	92 (20–237)	5.6 (1.6–7.8)	10 (2–20)	0.8 (0.3–1.4)

NOTE: The data between brackets refer to the minimum and maximum weighted average loads belonging to the refineries categorized by type of treatment. In some cases the results of treatment III would not appear to be better than those of treatment II. However, treatment III is applied nearly always where a more polluted effluent is involved and its general impact then is much greater.

Million Tons

Table 34. QUANTITIES OF LIQUID FUEL USED IN STATIONARY INSTALLATIONS
IN OECD COUNTRIES IN 1969[1]

LIQUID FUEL, INCLUDING LIQUEFIED GAS	OECD EUROPE	NORTH AMERICA	JAPAN
Refinery Fuel (Including refinery gas used as fuel)	30.3 (5.5%)	34.5 (5.9%)	4.4 (2.9%)
Thermo-electric plants	50.0	41.0	26.4
Industry	123.3	45.3	47.5
Other sectors (mainly domestic)	119.8	158.3	28.4

1. "OECD Statistics of Energy 1955-1969", Paris, 1971.
NOTE: The percentages between brackets are the percentages of refinery fuel relative to crude oil processed. The low percentage in Japan reflects a lower processing intensity, mainly topping operations.

VIII

INVESTMENT REQUIREMENTS FOR A CONTINUING[1] DEVELOPMENT

A. THE BACKGROUND AND FUNDAMENTAL ASPECTS OF OIL INVESTMENT

366.   During the decade ending in 1970, the world's capital and exploration expenditures for oil, excluding the USSR, other Eastern European countries and China, totalled $157 billion.   Expenditures of this magnitude were needed to maintain existing capacity and to supply a growth in demand of about 994 million tons for this period, representing an annual average growth rate of 7½ per cent.   After amortization, the gross investment in fixed assets employed by the oil industry at the end of 1970 amounted to $205.9 billion.   Traditionally, the proportion of capital needs raised on world money markets by international oil companies has been around 10 per cent of their new investment with the remaining 90 per cent derived from retained earnings, depreciation and depletion, and other non-cash charges. This picture, however, is changing as related by an international banking source[2].   They stated that in a group of US owned oil companies, invested capital increased by only 11.6 per cent compared with a 27.1 per cent rise for long term debt in the past 2 years.   Over the past 6 years, long term debt has increased from 12 per cent of capital employed to 20 per cent in 1970.

367.   During the decade ended in 1970 the oil industry's annual capital expenditures ranged from $10.5 billion in 1960 (dropping slightly in 1961 to about $10.4 billion) to $20.1 billion in 1970.   Each year subsequent to 1961 showed an increase with the biggest jump in 1968 of $2.3 billion, one year following the 1967 Suez crisis.

368.   Exploration expenses stayed around $1 billion annually until 1968 when they rose to nearly $1.4 billion annually.

369.   Table 35 shows the functional breakdown of the capital expenditures in the 1960 to 1970 period.

370.   In reviewing the decade of the sixties compared to the previous OECD report, "Oil Today - 1964", it is notable that there has been a

---

1. The dollar amounts shown in this chapter are in actual dollars.
2. 1970 Annual Financial Analysis of a Group of Petroleum Companies, the Chase Manhattan Bank.

153

decreasing percentage of investment made in oil production facilities. While the total figures are heavily influenced by the United States (60 per cent) compared to the other countries (28 per cent), the resultant 43 per cent for the sixties is less than the 56 per cent share shown in the previously cited report. Table 36 is a comparison of the two periods. The larger total of expenditure going to production reflects the more mature status of the production segment of the oil industry in the United States, compared with the more recently developing areas of other parts of the world. The decrease in both shares, however, for the later period indicates the requirement of more capital expenditure being made in the refining, marketing and chemical plant areas to satisfy the growing rate of demand for petroleum products throughout the world. This trend is continuing upward shown by the annual increases in expenditures over the period. The growth for the period in annual production expenditures of 29 per cent was far lower than the growth for refineries (254 per cent), chemical plants (231 per cent) and marketing (126 per cent).

371.   The functional breakdown of the $205.9 billion of gross fixed assets at year end 1970 is shown in Table 37.

B.   INVESTMENT IN THE UNITED STATES AND CANADA

372.   The capital expenditures in the domain of oil in the United States and Canada have developed an increasing trend since the last report. From $5.8 billion in 1960, the outlays rose to $9.5 billion in 1970. As noted earlier, the predominant characteristic of North American expenditures was the heavy emphasis on production. As reflected in the totals, production outlays rose steadily from $4.1 billion in 1960 to $5.2 billion in 1970. However, the share of capital going into production related to total capital expenditures has been decreasing, dropping from 71 per cent in 1960 to 54 per cent of aggregate capital expenditures in 1970. Chemicals and refining expenditures showed the largest percentage gains over the decade; however, marketing and pipelines were not far behind.

C.   INVESTMENT IN THE OECD EUROPEAN AREA

373.   OECD Europe capital outlays in 1970 were $3.26 billion, up 203 per cent over 1960, the high for the 11 year period. By far the bulk of the European outlays were in refining, marketing, and chemical plants (32, 28 and 22 per cent respectively) with production having only 11 per cent allocated to it. Refining has more than tripled expenditures while marketing has more then doubled during the 11 year period. Pipelines, however, have shown the greatest growth even though the share of the total in 1970 was less than 3 per cent.

D.   INVESTMENT IN THE MIDDLE EAST, AFRICA,
VENEZUELA AND THE FAR EAST

374.   Reflecting the comparatively low costs of obtaining production in the Middle East, total capital outlays for the period amounted to $5.5 bil-

154

lion. Production outlays accounted for 54 per cent of the total, refining 18 per cent and pipelines 14 per cent. A significant growth in pipeline expenditures occurred in 1969, accounting for 29 per cent of the capital outlays for that year, up 250 per cent over expenditures in 1960.

375.   African capital expenditures for the 1960-70 period totalled $7.05 billion and were fairly constant at around $600 million annually until the end of the 11 year period when they rose to around $800 million. Most of the increase went into production facilities which jumped about $200 million. In 1970 production accounted for almost 67 per cent of capital outlays reflecting development of Libyan and Nigerian fields. Pipelines and refining surpassed marketing for demands on capital expenditures.

376.   Venezuelan capital expenditure— 82 per cent for production— aggregated $2.4 billion for the 1960-70 period. In the 1953-62 period, the aggregate for that period was $4.7 billion. The peak year for capital expenditure was back in 1957 at $1.1 billion. Since that time the annual outlays have dropped to a low of $150 million in both 1966 and 1967. By 1969 they had risen to $340 million which is the high point for this period. In 1970 they dropped back to $300 million. In 1970 production accounted for three-fourths of the outlay with refining and pipeline accounting for 18 per cent.

377.   In the Far East capital expenditures totalled 11 billion for the 1960-70 period. Of this total one-half was expended for refineries and chemical plants, 28 per cent for marketing and only 15 per cent for production. This has been the pattern in the Far East for this period except for a $400 million increase for refining in 1970.

378.   Making a comparison of the amount of capital expenditures allocated to production to the amount of oil produced provides some interesting information. As expected, the more mature production in North America has required a proportionately larger capital outlay for production per barrel of oil produced because a very large number of wells produce a correspondingly small amount of oil per well. For example, in the United States production amounted to only 17 barrels per well per day (bpd). In Venezuela, however, per well production was 306 bpd. Table 40 depicts the relationship of capital expenditures to each barrel of oil produced in selected areas.

E. FUTURE INVESTMENT REQUIREMENTS AND HOW THEY CAN BE MET

379.   The continuing growth of oil demand will necessitate substantial capital investments throughout the world (excluding the USSR, other Eastern European countries and China). As stated previously, capital and exploration expenditures totalled $147 billion in the 1960s and were necessary to supply *a growth* in demand of 950 million tons. The *total* demand for this period, however, was for 13.1 billion tons. In the 1970s the demand for oil is expected to total nearly 26.8 billion tons (about 200 billion barrels). This future demand projection for the seventies is twice the demand required in the sixties.

155

380.   To supply this tremendous increase in volume will require enormous amounts of capital to be invested in the search for more oil and gas, to produce, transport, refine and market it. And there are no realistic prospects for any reduction in the average investment required per unit of petroleum moving to market. In truth, the opposite is much more likely to happen. In the sixties, the rate of capital spending fell below actual needs, particularly in exploration expenditures in the United States where spending rose only by 6.4 per cent for the 11 year period, from $625 million to $665 million, while all other capital expenditures were rising 59 per cent. The results of such a reduced rate of spending show that in 1960 there were 960 barrels of reserves added for each 1,000 barrels of production and had dropped to only 663 barrels of reserves added for each 1,000 barrels of production in 1969[1]. Therefore, this deficit trend needs to be made up by higher spending in the years to come.

381.   There are additional reasons why the unit cost of providing additional capacity is bound to rise in this decade. Where the search for new oil continues, an increasing proportion of it will be in a more hostile and costly environment such as Arctic areas, offshore regions, and at greater depths. Growing awareness of environmental conservation will require additional and costly manufacturing processes in refining as well as extra expenditures in marketing facilities and marine transportation protection. As more and more evidence is developed it appears that the industry will have to spend much more for costs associated with efforts to control the environment.

382.   Last, but certainly not least, is the realization of inflation and its impact on rising costs, particularly in new plant installations.

383.   In addition to the vast amount of money needed for capital and exploration expenditures, the petroleum industry will need approximately 50 per cent more for debt retirement, dividends and working capital.

384.   Therefore, the overall financial needs could very well be in excess of $550 billion, with the environmental costs mentioned earlier not included in this figure. Of this total, $365 billion will be spent in capital accounts and exploration. Dividends, retirement of debt and working capital will require about $185 billion. Regarding the effort to be placed in exploration it has been predicted that some $50 billion will be required to maintain adequate sufficiency, particularly in the United States. This is about five times the amount spent during the last ten years.

385.   The capital to meet these mammoth requirements can come from three potential sources: various provisions for capital recovery, net income and borrowed capital. In the past, capital recovery has ranked almost as important as net income as a source. Tax reform in recent years has reduced its effectiveness. For example, the Tax Reform Act of 1969 in the United States terminated the 7 per cent investment credit previously allowed, reduced the statutory depletion allowance rate from 27½ per cent

---

1. In 1970, there were 3,820 barrels of reserves added for each 1,000 barrels of oil produced, but 2,895 barrels were attributed to Prudhoe Bay, Alaska, discovered in 1968. Deleting Prudhoe Bay discoveries reduces the reserves added for each 1,000 barrels produced to 925 barrels about the same rate as prevailed in 1960.

to 22 per cent and precluded the use of excess foreign tax credits against non-mineral foreign income. These are the major actions reducing potential capital recovery. As a result, the industry will have to rely more and more on net income, outside borrowing and additional equity capital.

386.   Earlier it was estimated that some $365 billion would be needed in capital investment and exploration costs to meet a demand of almost 200 billion barrels. This is about $1.82 per barrel expenditure. Add to this working capital, dividends and debt retirement costs estimated at around $0.93 per barrel resulting in total expenditures anticipated at $2.75 per barrel.

387.   If this money were to be generated internally it would require the industry's net income to grow at a rate of 22 per cent a year. Such a rate would be almost three times as great as the industry's net income growth during the sixties. In fact, the trend has been the reverse. Eastern Hemisphere earnings of the group of the seven major companies, as estimated by the First National City Bank of New York[1], fell from 56.6 cents per barrel in 1960 to 33.0 cents per barrel in 1970. These earnings represent a return on net worth of 14.1 per cent in 1960 declining to 11.2 per cent in 1970. Maintaining the present rate of earnings would only provide some two-thirds of future capital requirements.

388.   The other source of capital is to borrow on the world money markets, where access for the oil industry depends on its relative position in terms of profitability to all other branches of industry. If the oil industry is to rely more heavily on the money market, some increase in revenue will have to be accomplished in order to improve its competitive position on that market.

389.   Whatever the source of financing —retained earnings or borrowing on the market— both will require a future level of revenue that includes the necessary margin for covering the cost of additional capital required to finance future investments. The question of spreading the capital requirements in time may imply a certain preference for a particular form of financing, which depends on the profit situation and individual appreciation of each company.

---

1. Energy Memo, October 1971, Petroleum Department, First National City Bank of New York.

Table 35. WORLD[1] CAPITAL EXPENDITURES IN THE DOMAIN OF OIL FOR THE PERIOD 1960-1970

In millions of dollars

	PRODUCTION	REFINERIES	MARKETING	MARINE	PIPELINES	CHEMICAL PLANTS	OTHER	TOTAL
United States:								
Dollars	44,435	6,720	10,300	640	3,340	5,360	2,680	73,475
Percent	60	9	14	1	5	7	4	100
Other World[1]:								
Dollars	22,610	17,710	14,325	14,750	4,515	5,805	2,110	81,825
Percent	28	22	17	18	6	7	2	100
Total:								
Dollars	67,045	24,430	24,625	15,390	7,855	11,165	4,790	155,300
Percent	43	16	16	10	5	7	3	100
1970 Total	7,230	4,000	3,220	2,615	850	1,525	685	20,125
1960 Total	5,610	1,130	1,425	1,110	550	460	240	10,525
Percent increase in annual expenditures over the period 1960 to 1970:	29	254	126	136	55	231	185	91

1. Excluding the USSR, other Eastern European countries and China.
SOURCE: Capital Investments of the World Petroleum Industry, 1970; The Chase Manhattan Bank.

158

Table 36. COMPARISON OF WORLD[1] SHARES IN CAPITAL EXPENDITURE IN THE DOMAIN OF OIL FOR THE (1953-1962) DECADE WITH THE (1960-1970) PERIOD

In percent of total investments

	PRODUCTION		REFINERIES		MARKETING		MARINE		PIPELINES		CHEMICAL PLANTS		OTHER		TOTAL	
	53-62	60-70	53-62	60-70	53-62	60-70	53-62	60-70	53-62	60-70	53-62	60-70	53-62	60-70	53-62	60-70
United States	71	60	10	9	8	14	1	1	5	5	3	7	2	4	100	100
Other World[1]	37	28	17	22	16	17	19	18	6	6	3	7	2	2	100	100
Total World[1]	56	43	13	16	12	16	9	10	5	5	3	7	2	3	100	100

1. Excluding the USSR, other Eastern European countries and China.

Table 37. WORLD[1] INVESTMENTS IN THE DOMAIN OF OIL IN GROSS FIXED ASSETS AS AT END YEAR 1970

In millions of dollars

GROSS	UNITED STATES		OTHER COUNTRIES[1]		WORLD[1]	
	DOLLARS	PERCENT	DOLLARS	PERCENT	DOLLARS	PERCENT
Production	54,700	56	26,625	25	81,325	40
Pipelines	6,525	7	7,000	6	13,525	7
Marine	1,175	1	21,265	20	24,440	11
Refineries	12,725	13	25,360	24	38,085	18
Chemical Plants	6,900	7	6,890	6	13,790	7
Marketing	12,600	13	18,785	17	31,385	15
Other	3,050	3	2,250	2	5,300	2
Total	97,675	100	108,175	100	205,850	100

1. Excluding the USSR, other Eastern European countries and China.

159

Table 38. CAPITAL DOLLARS SPENT IN 1970
IN PRODUCTION PER BARREL OF OIL PRODUCED

	Dollars per barrel	
	1960	1970
United States and Canada	1.49	1.30
Venezuela	0.22	0.17
Middle East	0.13	0.05
Africa	3.53	0.22
Far East	0.33	0.64

IX

NATURAL GAS

390.   Natural gas is a highly convenient fuel.   It is normally sulphur-free or readily purified and can be burned completely in simple appliances; it leaves no residue and does not give rise to atmospheric pollution to anything like the same extent as the other fossil fuels.   It is also a raw material for the manufacture of fertilizers and certain other chemical products.   The major disadvantage of gas is that its transport is relatively expensive.   To carry an equivalent amount of energy measured in BTUs, natural gas pipelines have to be about four times the size of oil pipelines; similarly, a tanker to transport LNG has to be perhaps twice the size of an equivalent oil tanker, and construction costs of an LNG tanker are high, as indeed are those of the related liquefaction, storage, and gasification.   Thus, transport costs have a significant influence on the delivered price of natural gas whether by sea or by pipeline, and a gas field in an area remote from market may for this reason be of less economic value than for those fields more suitably located.

391.   About one-third of the world's hydrocarbon reserves (excluding tar sands and shale) are in the form of natural gas, either in isolation or in association with oil.   Natural gas provides about one-fifth of current world energy supplies; consumption of gas in 1971 approached 1,100 billion cubic metres, equivalent to some 950 million tons of oil.   Trends in the consumption of gas during the period 1960-70, world-wide and for the OECD area, are shown in Table 1 of Chapter I.

392.   The impact of natural gas on the consumption of energy in the OECD European Member countries was the subject of a detailed study, published in 1969, by a joint working party of the OECD Energy Committee and the Oil Committee[1].   This report, which also included comprehensive accounts of the development of the natural gas industry in the United States and Canada, and of likely availabilities in areas peripheral to the OECD European area, is still valid in its general approach to the subject and no attempt will be made here to review its findings.   Rather we shall limit our coverage to certain aspects which now need revision.

393.   Natural gas resources in the OECD European area were placed in the 1969 study at 4.1 trillion cubic metres, of which nearly 3.3 trillion cubic metres were considered as "proved recoverable".   These estimates reflected the position at about mid-1968 and may be compared with an estimate of proved world reserves at that time of 38 trillion cubic metres.   Present estimates of proved reserves are higher; a world total of 45 trillion cubic metres of which 4.6 trillion cubic metres are in the OECD European area.

---

1.   "Impact of Natural Gas on the Consumption of Energy", OECD, Paris, 1969.

394.   Natural gas is the most rapidly expanding sector in OECD European energy supply, and forecasts of demand are therefore subject to relatively wide margins of error.   Indeed the range of estimates given in the 1969 report of Europe's likely demand in 1980 may now be considered as pessimistic with the then upper range of 195 Mtoe now falling below the more likely demand now estimated to be 231 Mtoe.   Two complementary factors explain this upward shift.   First, the level of total energy demand is expected to be higher in 1980 than was foreseen in the report and, secondly, production levels of natural gas are expected to be higher, notably in the Netherlands and Germany.   Overall, a demand level of 231 Mtoe for natural gas in Europe would imply a 13 per cent share of total energy demand in 1980.

395.   Table 1 in summary fashion, and Tables 2, 4, 6 and 8 of Chapter I in more detail, illustrate the expected development of natural gas in the main OECD areas (Europe, North America and Japan) and in the rest of the world during the 1970s.   World-wide, natural gas is likely to increase its share of world energy demand from 19 per cent in 1970 to some 20 per cent in 1980, thus enlarging only slightly its relative contribution.   North America is likely to remain the premier gas consuming area of the world, with some 30 per cent of its total energy requirements provided by gas (if supply difficulties can be overcome).

396.   Trade in natural gas by pipelines and in liquefied form by methane carrier is of growing importance in Member country energy economies.   The improving technology for transporting liquefied natural gas (LNG) is of special interest since it represents the only practical method of transporting gas where convenient land routes are not available.   Substantially all the LNG now moving in world trade is consumed within the OECD area and will continue to be so in the years ahead, but with a change in geographical emphasis compared with the present scene.   The main areas of development of LNG imports in the next decade are likely to be the United States (Atlantic Seaboard and possibly Southern California) and Japan, and by further quantities from North Africa to Southern Europe.   US imports of LNG, as with Europe, are to originate in North and West Africa.   There are prospects for both the US and Japan of importing LNG from the Soviet Union.   Brunei will contribute substantially to Japan's LNG supplies.   An illustration of the present position, and likely trend in LNG trade is set out in the following tabulation:

LNG SUPPLIES

Billion cubic metres.

	1970 Actual	1980 Forecast
OECD Europe	1.5	8.5 - 10
United States	0	20 - 30
Japan	1.5	10 - 20

NOTE.   The 1980 estimate for Europe is based on current contracts with Algeria and Libya.   The estimate for Japan reflects existing contracts on the lower end and the possibility of enlargement of LNG supplies from other sources during the decade.

X

OIL PRODUCT PRICES

A. TRENDS DURING 1960-1969

397.   During the 1960s, the prices (excluding taxes) for the major oil products outside North America and the Communist area exhibited a declining trend.   Over this period pre-tax prices of gasoline fell by more than 10 per cent, gas/diesel-oil prices by about 20 per cent and prices of heavy fuel-oil by about 15 per cent.   Figures 11, 12 and 13 show the course of gas/diesel-oil, fuel-oil and motor gasoline prices in selected major cities on the European continent during the past decade while figures 14 and 15 show those for the United States and figures 16, 17 and 18 those for Japan.

398.   This declining trend was opposite to the course of prices for many other commodities and for wage rates during the same period.   For instance, when setting 1960 prices at a base of 100, the indices in 1970 for the United Kingdom were as follows: for heavy fuel-oil: 85; for all manufactures: 135; for coal: 140; and for hourly wage rates: 170.

399.   When seeking an explanation for the phenomenon of declining oil product prices in an inflationary economy, the following factors must be taken into consideration.

a)   Notwithstanding the general effects of inflation, refining and distribution costs remained fairly constant as a result of improvements in the efficiency of operations, technological progress and the benefits of the economies of scale.   The increase in the size of refineries, mentioned in Chapter V, paragraph 221, contributed to the latter.

b)   The introduction of tankers of the 200,000 dwt class (see Chapter VI, A) and the increasing importance of North and West Africa as a supplier of shorthaul crude oil for Europe, brought about a reduction in freight costs.

c)   The rapid development of new oil resources in Africa and the Middle East, combined with the continued import restrictions in force for the United States, gave rise to a potential surplus of crude oil.   Part of the new oil was in the hands of newcomers to oil production.   These new supplies added extra pressure in an already highly competitive market.   The increased competition which these newcomers brought to the market, coupled with the depressing impact of surplus oil in general, not only found expression in the prices, but also in the profit margins of the interna-

163

tional oil companies. To illustrate, Eastern Hemisphere earnings of the seven largest oil companies reportedly declined from 56.5 cents a barrel in 1960 to 32.7 cents in 1970[1]. This development was to some extent countered by an increase in government take of Middle East and African crude oil from 71 cents a barrel in 1960 to 86 cents in 1970. The US market is an isolated market and the development of oil product prices in the United States has been quite different from that in other parts of the world. Gasoline prices (excluding taxes) remained fairly stable during the first half of the 1960s, but then increased by more than 15 per cent from 1965 through 1970. According to the Independent Petroleum Association of America, average wholesale prices for fuel-oil in 1970 were also higher than in 1960. For light fuel an increase of more than 15 per cent took place in the second half of the 1960s. Prices of heavy fuel generally declined for most of the decade. However, an increase in 1970 more than compensated for the decline in the earlier years. The 1970 price level for this fuel was more than 20 per cent above the 1960 level, reflecting in part the substantial— and somewhat unanticipated— growth in demand for this fuel.

400.   Fuel sales are now looked to as an important source of government revenue and over the years the various forms of taxes which have been built into the retail price for gasoline and the like now have become the dominant element in the final price paid by the consumer. This is particularly so for Europe, where, in the European Community for instance, taxes at the beginning of 1971 represented between 62 per cent and 77 per cent of the retail price[2] for higher grade gasoline, and ranged between 11 and 20 US cents per litre sold; in the United States, federal and state taxes on the average amount to about 11 cents per gallon (about 3 cents per litre) or some 32 per cent of the total, and are identical to the quoted Gulf Coast Cargo price. Figure 19 illustrates the relative importance of the individual elements in retail gasoline prices in the United States.

B.   DEVELOPMENTS IN 1970 AND 1971

401.   In 1970, for a variety of reasons mentioned in Chapter IV A, the buyers' market of the 1960s was turned into a sellers' market and host governments were, under the circumstances, able to secure an increase in their take.

402.   In 1971 the host governments were able to obtain further increases in their take through the Teheran and related agreements which together with the earlier rises meant substantial increases in the tax paid costs for all producing companies. The increases varied from about 40 cents per barrel on long haul crude from the Middle East (an increase of 40 per cent) to 79 cents in Nigeria and 90 cents in Libya (increases of 60 to 70 per cent in tax paid costs). At the same time, the Venezuelan Government had

1. First National City Bank.
2. Prices at 1st January 1971. See "Energy Statistics", EEC Yearbook, 1971.

imposed increased take obligations which raised costs by about 47 cents per barrel and Algeria had adjusted its terms to increase its revenue by 90 cents per barrel.

403.   Oil product prices rose towards the end of 1970 reflecting the initial take increases and a sharp increase in tanker freight rates.   These higher prices were maintained through the earlier part of 1971 on account of the further take increases despite declining freight rates; in the second half of 1971 much of the gain in product prices disappeared because of a slackening of growth rate and the continuing decline in freight rates.

C.   FUTURE TRENDS

404.   There are various reasons which make it increasingly unlikely that the stability in oil product prices which Western Europe and Japan experienced during the 1960s will be continued during the 1970s.   A number of factors lead us to believe that product prices can be expected to increase. The most important of these are:

a)   The Teheran and related agreements provide for annual increases in the posted prices up to 1975 with the result that the tax-paid costs of long-haul Middle East crudes alone will be increased by about 6.5 cents in each of the years 1973-1975.

b)   Whereas during the 1960s the United States was largely self-sufficient in oil, the expectations for this decade are quite different and the US is expected to become increasingly dependent on imports to meet its growing demand, which imports by the early 1980s will probably have reached a level of Western Europe's oil imports today.   In the past the US has looked to Canada and Venezuela for its oil supplies but there is reason to believe that these countries will not be able to supply oil at a pace in keeping with the expanding needs of the US.   Consequently the Eastern Hemisphere will increasingly be called upon to supply growing amounts of oil to that country.

c)   The amount of oil which will have to be found in the next 20 years to meet demand means that the search for oil must take place in more hostile or difficult environments, for example, the deep sea where exploration and production costs are 3 to 10 times those on land, which can only lead to ever higher costs— and prices— for the oil.

d)   As pointed out before, one of the major contributing factors to the stability of oil prices during the 1960s was the ability of the oil industry to carry through operational economies in practically all phases of its operations, but in particular in tanker transportation and refining.   It now seems that, at least for the time being, the opportunities for further economies have been practically exhausted, so that inflationary effects, which thus far have been compensated for by improvements in efficiency, will in the future have to be reflected in prices.

e)   Another element of additional costs arises from the need to provide adequate safeguards against pollution.   Up to 10 per cent of any new refinery investment is for antipollution protection.

Moreover, the need to preserve the environment asks for changes in product requirements from refineries. The process of removal of sulphur from residual fractions is very costly, as is the manufacture of unleaded gasoline.

f) In June, 1971 the Council of the OECD recommended to the governments of the European Member countries that a stock level of at least 90 days average inland consumption of the previous calendar year should be achieved as soon as possible. The capital outlay required to raise 1970 average stock levels in OECD Europe to 90 days is estimated at roughly $1,000 million. An incremental effect of the additional costs of storage on product prises seems likely.

g) The financial needs of the oil industry are higher than ever before. As already stated in Chapter VIII, paragraph 384, these needs could very well be in excess of $550 billion in the 1970s. Finding the necessary capital will present major problems for the oil industry as there will be a gap of 15 to 20 per cent relative to the capital needs for the decade, which will most likely have to be covered by increasing revenues.

166

Figure 11

DEVELOPMENT OF PRICES FOR GAS/DIESEL OIL AT SELECTED PLACES
OF THE EUROPEAN COMMUNITY, (EXCEPT TAXES AND DUTIES)

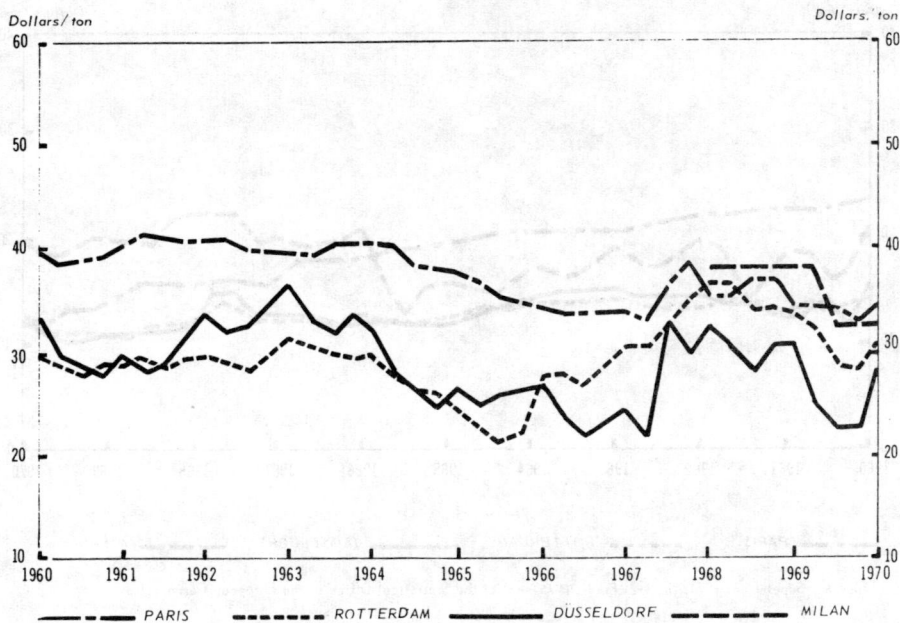

Dollars/ton

Dollars. ton

60 — 60

50 — 50

40 — 40

30 — 30

20 — 20

10 — 10

1960 1961 1962 1963 1964 1965 1966 1967 1968 1969 1970

— — — PARIS ━━━━ ROTTERDAM ━━━━ DÜSSELDORF ━ ━ ━ MILAN

Source : Supplement 1971, Nr. 1-2 *Fuel Oil Prices* of the Statistical Office of the European Communities.

DEVELOPMENT OF PRICES FOR HEAVY FUEL OIL AT SELECTED PLACES
OF THE EUROPEAN COMMUNITY, (EXCEPT TAXES AND DUTIES)

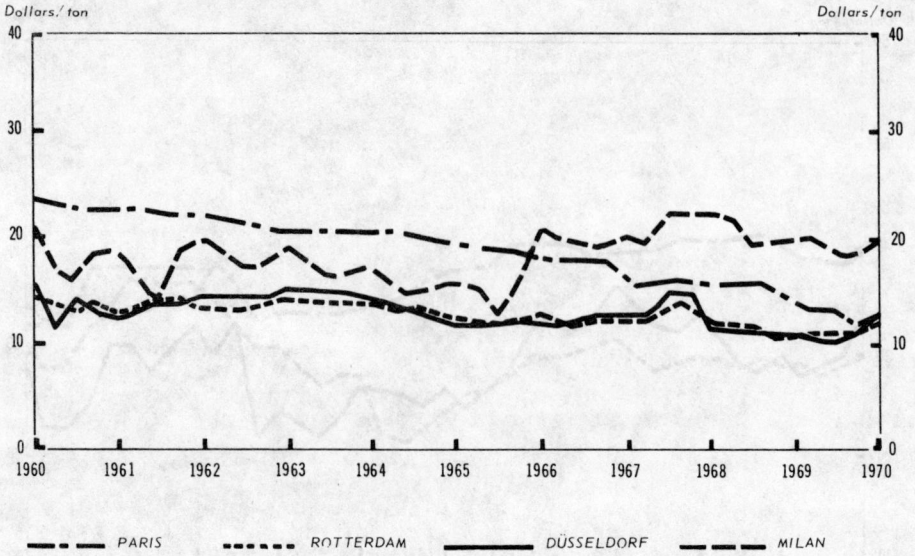

Dollars/ton

Dollars/ton

— · — PARIS   · · · · · · ROTTERDAM   ———— DÜSSELDORF   — — — MILAN

Source : Supplement 1971, Nr. 1-2 *Fuel Oil Prices* of the Statistical Office of the European Communities.

168

Figure 13

DEVELOPMENT OF PRICES FOR MOTOR GASOLINE (PREMIUM) IN SEVERAL COUNTRIES
OF THE EUROPEAN COMMUNITY (EXCEPT TAXES AND DUTIES)

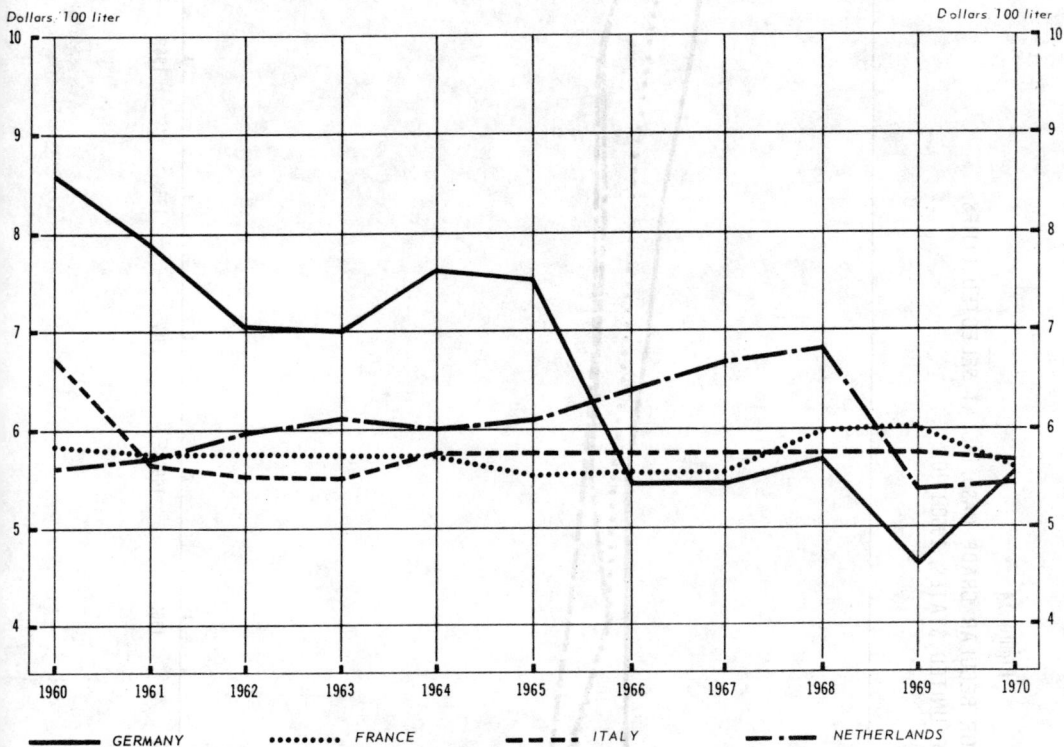

Source : Yearbook 1960-1970 *Energy Statistics* of the Statistical Office of the European Communities.

Figure 14

DEVELOPMENT OF PRICES* FOR REGULAR GRADE GASOLINE AT SELECTED CITIES
IN THE UNITED STATES, 1960-1970

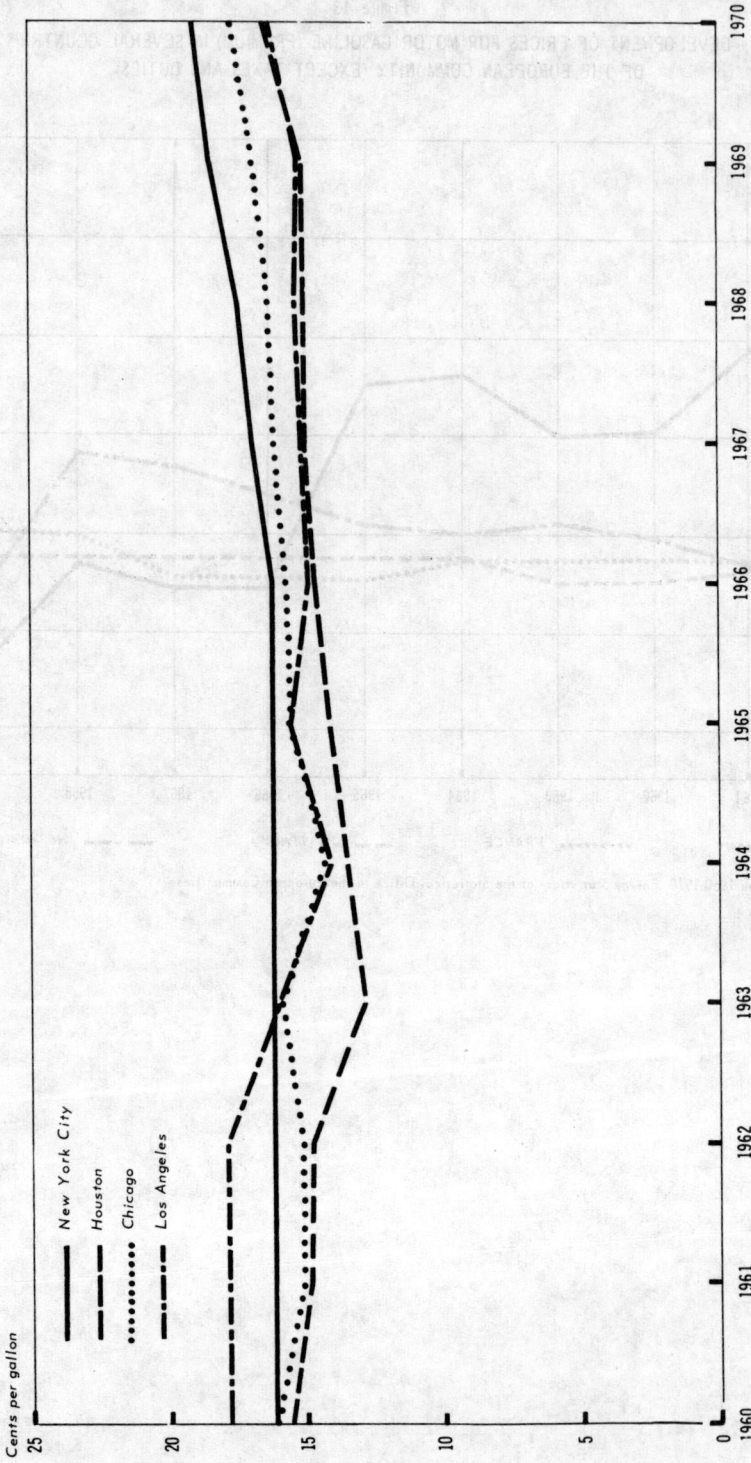

Cents per gallon

New York City
Houston
Chicago
Los Angeles

25
20
15
10
5
0

1960 1961 1962 1963 1964 1965 1966 1967 1968 1969 1970

* Dealer tank wagon price, excluding taxes.

Source : API Fact and Figures, 1971 Edition; Platt's Pricing Handbook.

170

DEVELOPMENT OF AVERAGE REFINERY PRICES FOR No 6 FUEL OIL * AT
SELECTED CITIES IN THE UNITED STATES, 1960-1970

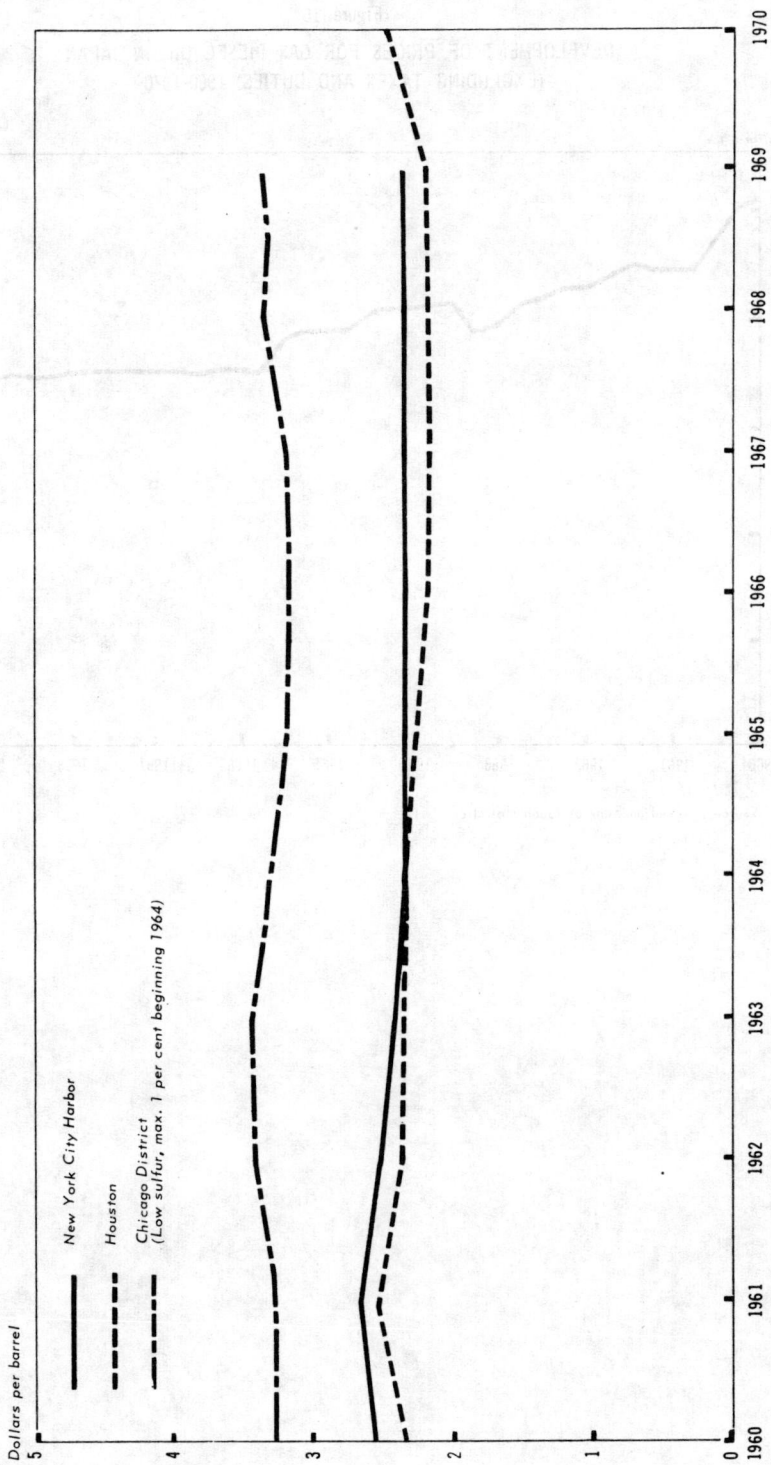

Dollars per barrel

New York City Harbor

Houston

Chicago District
(Low sulfur, max. 1 per cent beginning 1964)

5

4

3

2

1

0

1960 1961 1962 1963 1964 1965 1966 1967 1968 1969 1970

* No sulfur guarantee unless otherwise noted.

Sources : API Facts and Figures, 1971 Edition; Platt's Pricing Handbook.

Figure 16

DEVELOPMENT OF PRICES FOR GAS/DIESEL OIL IN JAPAN
(EXCLUDING TAXES AND DUTIES) 1960-1970

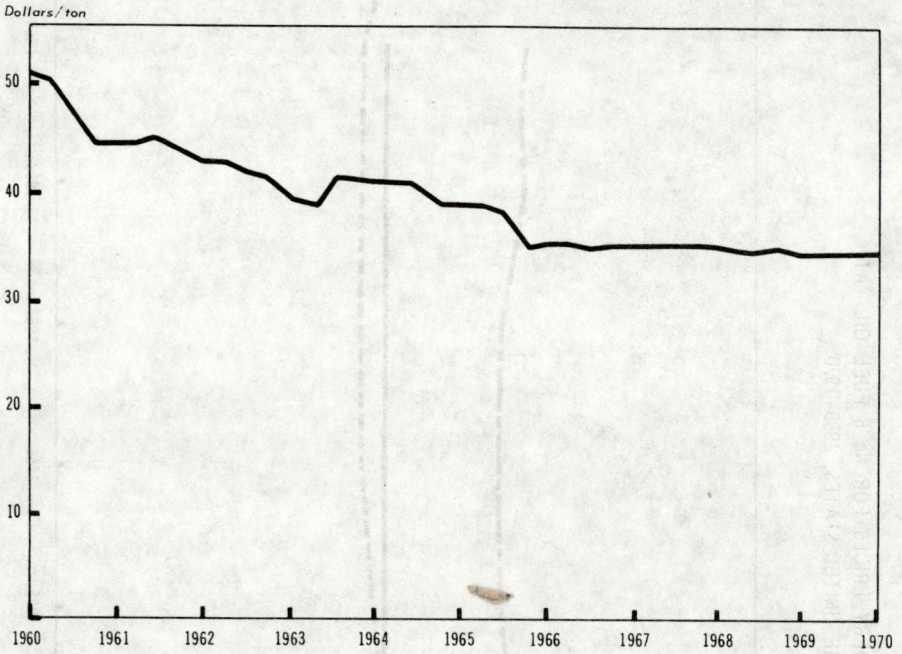

Source : Based on Bank of Japan statistics.

Figure 17

DEVELOPMENT OF PRICES FOR HEAVY FUEL OIL IN JAPAN
(EXCLUDING TAXES AND DUTIES) 1960-1970

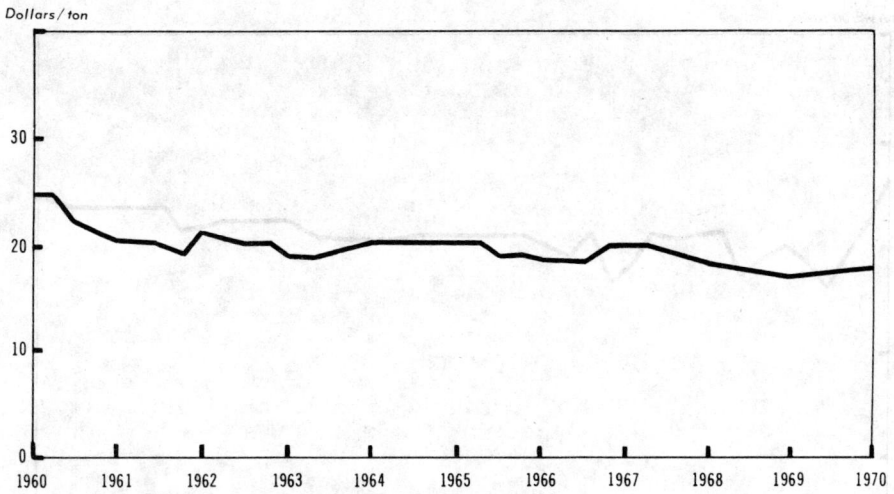

Source : Based on Bank of Japan statistics

Figure 18

DEVELOPMENT OF PRICES FOR MOTOR GASOLINE IN JAPAN
(EXCLUDING TAXES AND DUTIES) 1960–1970

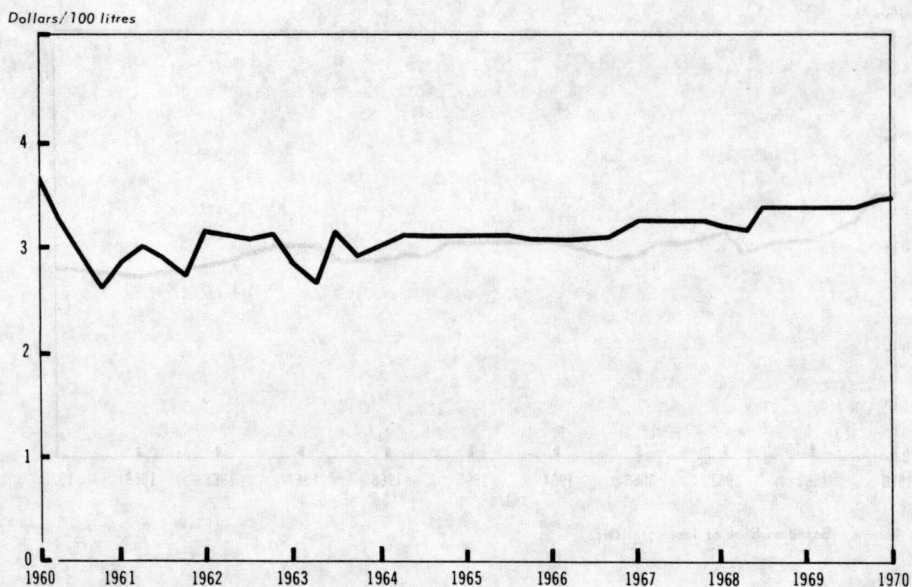

Source : Based on Bank of Japan statistics.

174

Figure 19
HOW THE PRICE OF A GALLON OF GASOLINE IS MADE UP
IN THE UNITED STATES

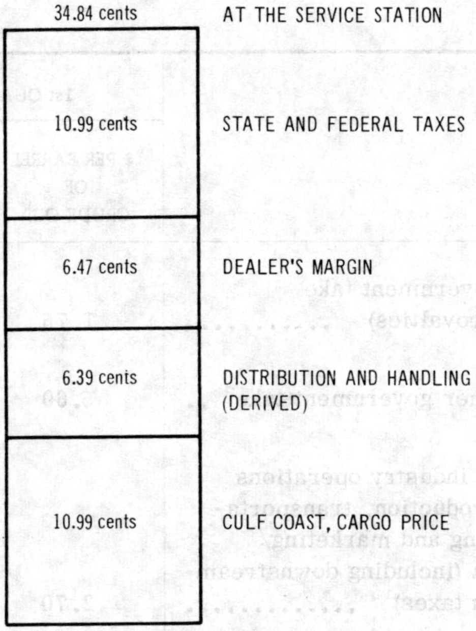

34.84 cents	AT THE SERVICE STATION
10.99 cents	STATE AND FEDERAL TAXES
6.47 cents	DEALER'S MARGIN
6.39 cents	DISTRIBUTION AND HANDLING (DERIVED)
10.99 cents	CULF COAST, CARGO PRICE

Average US Regular Grade Gasoline, 1969.

Sources : API *Petroleum Facts and Figures*, 1971 Edition, Washington, 1971 ;
Platt's *Oil Price Handbook*, 1969 Prices.

Table 39.   BREAKDOWN OF AN AVERAGE PRICE OF A BARREL AND OF A TON OF CRUDE OIL AS PAID BY THE CONSUMER IN SELECTED WESTERN EUROPEAN COUNTRIES [1]

	1st QUARTER 1972	
	($ PER BARREL OF CRUDE OIL)	($ PER TON OF CRUDE OIL)
Average host government take (taxes and royalties)	1.75	12.25
Average consumer government take[2] ..	5.60	39.20
Average cost of industry operations covering production, transportation, refining and marketing/ distribution (including downstream corporation taxes)	2.70	18.90
Average industry margin (for reinvestment and distribution to shareholders)	0.35	2.45
Weighted average gross proceeds per barrel and ton of crude	10.40	72.80

1. Countries included: United Kingdom, Germany, France, Italy, Benelux, Sweden.
2. This item does not, in reality, consist of duties charged on energy as such. A large proportion of this form of taxation represents a recoupment of financial burdens assumed by the community at large for the benefit of road transport.

Explanatory Note: All figures are derived from country weighted averages; they have been calculated from the 1st quarter of 1972 at the then current prices, on the basis of a close approximation to actual geographical mix of crude imports from main supply sources plus transportation cost at L(arge) R(ange) 2 AFRA, including inland pipeline transportation cost in the case of Germany.

Annex I

COUNTRY CHAPTERS

CONTENTS

Australia 179
Austria 185
Belgium 188
Canada 191
Denmark 196
Finland 197
France 200
Germany, Federal Republic of 206
Greece 211
Iceland 214
Ireland 215
Italy 217
Japan 220
Luxembourg 225
Netherlands 227
Norway 231
Portugal 234
Spain 237
Sweden 239
Switzerland 242
Turkey 247
United Kingdom 249
United States 252

*
* *

The monographs which make up this Annex were prepared by the Member countries concerned.

The statistics contained in the individual Country Chapters may not be identical with the figures given in the body of the Report and in Annex II which have been derived from a base which is homogeneous between countries.

AUSTRALIA

Principal Oil Statistics

Million metric tons

	1966	1967	1968	1969	1970
Crude and process oils					
Indigenous production	0.4	1.0	1.8	2.1	8.1
Imports	17.9	19.5	20.0	20.2	15.9
Total supply	18.3	20.5	21.8	22.3	24.0
Refinery intake	18.2	20.3	22.0	22.2	23.8
Exports	-	-	-	-	0.1
Total disposals	18.2	20.3	22.0	22.2	23.9
Petroleum products					
Refinery output	16.1	17.9	19.6	19.7	21.1
Output from other sources ...	-	-	-	-	-
Imports	1.1	0.9	0.9	2.3	2.6
Total supply	17.2	18.8	20.5	22.0	23.7
Deliveries to inland consumption	14.3	15.5	17.1	19.0	19.9
Bunkers	1.8	2.1	2.3	2.2	2.3
Exports	1.1	1.1	1.1	0.8	1.4
Total disposals	17.2	18.7	20.5	22.0	23.6
Refinery capacity					
Crude distillation capacity at end year	23.0	25.1	27.5	28.0	30.5

179

INDIGENOUS PRODUCTION AND IMPORTS OF CRUDE OIL

Production of Indigenous Crude Oil

The currently producing Australian fields are: Moonie and Alton in Queensland (about 117 thousand tons per year), which commenced production in February 1964; Barrow Island in Western Australia (about 2.3 million tons per year), which commenced production in March 1967; and Bass Strait (about 14.1 million tons per year).

Of the Bass Strait fields Barracouta commenced production in October 1969 and Halibut in March 1970. Kingfish commenced production in the second quarter of 1971, and will achieve full production in the first quarter of 1972, with the completion of the second offshore platform and of other facilities. Production from Bass Strait is expected to reach the rate of 15.8 million tons per year in the first quarter of 1972. With the 2.5 million tons per year expected from Barrow and Moonie, this would mean that Australia would be producing about 66 per cent of its requirements of crude by the beginning of 1972.

Quality of Indigenous Crude Oil

All indigenous crude oil so far discovered has been of a light type, consisting mainly of fractions from which motor spirit, kerosene, aviation turbine fuel and diesel-oils are produced.

Australian crudes are deficient in heavy ends, so that the local requirements for fuel-oil, lubricating oil and bitumen cannot be met from this type of crude oil.

Government's Oil Absorption Policy

In September 1965, the Government affirmed its policy that as far as possible local refineries use the crude oil produced in Australia. The Prime Minister, in October 1968, once more reaffirmed this policy, announcing that for 10 years, beginning 18th September, 1970, refineries in Australia are to be required to process Australian crude oil to provide as far as possible the full requirements of the Australian market for petroleum products.

Imports

Imports of crude oil into Australia are necessary to provide the heavier products, in which Australian crudes are deficient. In some cases topped crudes are processed to provide the necessary quantities of heavy fuel-oil, lubricating oils and bitumen. These imports can be expected to continue until a suitable heavy crude is discovered in Australia. In 1970, 66 per cent of Australia's crude oil requirements were imported, (see Provisional Oil Statistics), but this is expected to fall to 40 per cent in 1971 and 35 per cent in 1972, as local production increases.

DEVELOPMENT OF REFINERY CAPACITY

Prior to 1954 there were only three small refineries in Australia. Since then, with Government encouragement, oil companies have constructed refining facilities to the extent that today ample refining capacity is available to meet the whole Australian demand except in the case of a few specialised products.

There are ten modern refineries in Australia containing, in addition to atmospheric and vacuum distillation, various combinations of the following processes: reforming, catalytic cracking, hydrocracking, alkylation, polymerisation and desulphurisation. The total crude oil distillation capacity is currently 30.5 million tons per year. Lubricating oil plants are integrated at three of the refineries; production of lubricating oil stock in 1970 was 367,000 t. Bitumen production in 1970 was 447,000 t. Capital investment in the refining of petroleum to mid-1970 was A$570 million.

GROWTH OF INLAND CONSUMPTION OF MAIN OIL PRODUCTS

The total annual consumption of petroleum products for inland fuel usage has risen from 10.1 million tons in 1960 to 20 million tons in 1970. Inland consumption is expected to reach 37.5 million tons in 1980.

In 1960 motor spirit had by far the major share of the inland[1] petroleum market, with 47 per cent, followed by furnace fuel with 18 per cent, then automotive distillate with 9.5 per cent and industrial diesel oil with 8 per cent.

By 1970 the pattern had changed, and had been set for the next decade. Motor spirit now has 37 per cent of the inland market and this is expected to decrease to 31 per cent by 1980. Furnace fuel however has now obtained 25 per cent, and is expected to increase slightly, until by 1980 it will represent 27 per cent of the inland petroleum market. Consumption of automotive distillate more than trebled between 1960 and 1970, and is expected to increase a further 2.3 times within the next ten years, when it will hold 15 per cent of the inland petroleum energy market.

The other significant increase in consumption is of aviation turbine fuel which held 3 per cent of the total inland petroleum market in 1960 and now has 5 per cent. It is expected to obtain 7 per cent in 1980.

The consumption of industrial diesel oil is not expected to change significantly in the next decade. Its share is seen as declining from a current 4 per cent to 2 per cent in 1980.

The advent of natural gas, especially in the last year, has to some extent curtailed the previously projected rise in consumption of furnace fuel.

OIL CONSUMPTION IN RELATION TO TOTAL ENERGY CONSUMPTION

Total energy consumption in Australia has risen from 295×10^{12} kCal in 1960 to 508×10^{12} kCal in 1970. The estimated total energy consumption in 1980 is 940×10^{12} kCal. The average annual growth rates during these two decades are 5.6 per cent and 6.4 per cent respectively. The estimated growth rate for the decade to 1980 fluctuates between 5 per cent and 7.2 per cent per year. An 88 per cent increase in total energy consumption is expected over this period.

The total consumption of petroleum products showed an average annual growth rate of 8.4 per cent between 1960 and 1970, and the estimated

---

1. The inland petroleum market includes all deliveries in Australia except those for use as ships bunkers on seagoing vessels.

annual growth rate for the next decade is 6.4 per cent, giving an increase in consumption of 86 per cent over the 1970 figures. The advent of natural gas will curtail the high growth rates experienced in the 1960s.

In 1960 petroleum products held 37 per cent of the total energy market compared with 55 per cent held by black and brown coal. By 1970 this had risen to 50 per cent (Coal 43 per cent), and it is expected that this share will be maintained until 1980, with that of coal declining slightly to 38 per cent.

The principal Australian oil statistics for the last five years are seen in the table which precedes this section.

AUSTRALIA

Principal Natural Gas Statistics

Billion[1] cubic metres
at 15°C and 760 mm Hg.

	1966	1967	1968	1969	1970
Indigenous production[2]	0.004	0.004	0.003	0.227	1.295
Imports	-	-	-	-	-
Total supply	0.004	0.004	0.003	0.227	1.295
Inland consumption	0.004	0.004	0.003	0.227	1.295
Exports	-	-	-	-	-
Total disposals	0.004	0.004	0.003	0.227	1.295

1. Billion as used in this Report is 10^9.
2. Excluding waste and flared gas, processing plant fuel, losses and return to formation.

INDIGENOUS PRODUCTION AND IMPORTS OF NATURAL GAS

The production of natural gas in Australia is currently taking place in Queensland, Victoria and South Australia.

In Queensland, the gas reserves of the Roma area fields are committeed to supply 2.8×10^9 m³ over the next 15 years commencing in 1969 for industrial and domestic use in the Brisbane area. The pipeline to Brisbane with a total length of 450 km., commenced operating in March 1969.

In Victoria, the gas is supplied from two offshore fields. Production from the first field commenced in March 1969. The pipeline from this field to the State capital (Melbourne) is approximately 260 km. long.

In South Australia natural gas production from fields in the Cooper Basin commenced in late 1969. The gas is transmitted through a 780 km. line to the State capital and other places in South Australia for industrial and domestic consumption. In 1970 some 0.651×10^9 m³ of pipeline quality gas were produced for sale and field use.

A preliminary agreement has been signed for the supply of gas to New South Wales from South Australia through a proposed 1,320 km. pipeline provided that reserves of at least 57×10^9 m³ of gas are proved within the next year.

Western Australia has reserves of at least 12×10^9 m³. A 410 km. pipeline to Perth is scheduled for completion in late 1971.

183

There are neither imports nor exports of natural gas in Australia at the present time.

INLAND CONSUMPTION OF NATURAL GAS

Victoria commenced using natural gas in late 1969 for both domestic and industrial sectors of the market. Domestic conversion to natural gas was completed early in 1971. A new natural gas fuelled power station is expected to commence operating in 1976. Sales of natural gas to industrial consumers have been very successful.

The major consumer of natural gas in South Australia is the Electricity Trust of South Australia which uses the gas as fuel for electric power generation. The South Australian Gas Company reticulates natural gas in the Adelaide region and two cement factories are large consumers of natural gas as a fuel.

In Queensland, the bulk of the gas transported from the fields is used as a feedstock and fuel for the manufacture of nitrogenous fertilizers. Two towns on the pipeline route are supplied with natural gas which is also used as fuel for electricity generation at the source of the pipeline.

Natural gas should be available to consumers in Western Australia during 1971. The major consumer is expected to be the alumina industry. The State Electricity Commission of Western Australia will reticulate the natural gas in Perth.

Australian consumption of natural gas in 1970 was 1.3×10^9 m³. This is expected to reach 10.1×10^9 m³ in 1980.

CONSUMPTION OF NATURAL GAS IN RELATION TO TOTAL ENERGY

From 1964 to early 1969 the only consumption of natural gas was by a small power station in Queensland. In 1970 Australian consumption was 11.5×10^{12} kCals, which constituted 2.3 per cent of the total energy consumed. At this early stage of the introduction of natural gas it is very difficult to assess future consumption levels. It is estimated that by 1980 natural gas consumption will be 89×10^{12} kCals, representing 9.2 per cent of the total primary energy consumption in Australia. The average annual growth rate over this period is 28.3 per cent. From 1970, natural gas will be the third major contributor to the total energy market after oil and coal; it consolidates this position throughout the decade.

Principal Australian natural gas statistics can be seen in the table which precedes this section.

AUSTRIA

Principal Oil Statistics

Million metric tons

	1966	1967	1968	1969	1970
Crude and process oils					
Indigenous production	2.76	2.68	2.72	2.88	2.80
Imports	1.58	1.59	2.32	2.46	3.57
Total supply	4.34	4.27	5.04	5.34	6.37
Refinery intake	4.18	4.51	4.99	5.37	6.20
Exports	–	–	–	–	–
Total disposals	4.18	4.51	4.99	5.37	6.20
Petroleum products					
Refinery output	4.01	4.34	4.79	5.13	6.00
Output from other sources ...	–	–	–	–	–
Imports	2.13	2.14	2.68	3.00	3.27
Total supply	6.14	6.48	7.47	8.13	9.27
Deliveries to inland consumption	5.87	6.39	7.32	8.03	8.96
Bunkers	–	–	–	–	–
Exports	0.10	0.13	0.15	0.18	0.18
Total disposals	5.97	6.52	7.47	8.21	9.14
Refinery Capacity					
Crude distillation capacity at end year	4.66	4.66	4.66	4.72	8.20

185

INDIGENOUS PRODUCTION AND IMPORTS OF CRUDE OIL

Indigenous production of crude oil is limited by the extent of the reserves currently available. No considerable increase in production is to be expected. Imports of crude are rising steeply as as result of growing demand, following the expansion of refinery capacity. This trend is likely to continue in the next few years.

DEVELOPMENT OF REFINERY CAPACITY

Refinery capacity was increased and in fact almost doubled in 1970. Between now and 1974, capacity is to be expanded to 11 million tons a year.

GROWTH IN INLAND CONSUMPTION OF MAIN OIL PRODUCTS

The trend of inland consumption of oil products is as follows:

	1966	1967	1968	1969	1970	1975
Gasoline	1.13	1.24	1.35	1.44	1.58	2.30
Gas/diesel oil	0.93	1.02	1.11	1.24	1.62	2.90
Fuel-oil	2.99	3.29	3.96	4.29	4.50	6.00

OIL CONSUMPTION IN RELATION TO TOTAL ENERGY CONSUMPTION

Oil consumption as a percentage of the total supply of energy (expressed in coal equivalent) was:

	1966	1967	1968	1969	1970
	34.5	37.6	40.4	42.2	42.3

AUSTRIA

Principal Natural Gas Statistics

Billion[1] cubic metres
at 15°C and 760 mm Hg.

	1966	1967	1968	1969	1970
Indigenous production[2]	1.87	1.80	1.63	1.48	1.90
Imports	-	-	0.26	0.82	0.92
Total supply	1.87	1.80	1.89	2.30	2.82
Inland consumption	1.87	1.80	1.89	2.30	2.82
Exports	-	-	-	-	-
Total disposals	1.87	1.80	1.89	2.30	2.82

1. Billion as used in this Report is 10^9.
2. Excluding waste and flared gas, processing plant fuel, losses and return to formation.

INDIGENOUS PRODUCTION AND IMPORTS OF NATURAL GAS

A small increase in the production of natural gas appears possible, but this will be restricted by the size of the reserves at present available. The growing demand means an increase in imports, and these will rise still more in the future.

INLAND CONSUMPTION OF NATURAL GAS

Inland consumption also appears likely to grow in future. Between 1966 and 1970, consumption in the main consumer groups rose as follows:

Industry .... from 710 million to 966 million m³ at NTP*
Power stations .... from 421 million to 818 million m³ at NTP
Gasworks .... from 368 million to 478 million m³ at NTP
Petrochemicals .... from 83 million to 131 million m³ at NTP

CONSUMPTION OF NATURAL GAS IN RELATION TO TOTAL ENERGY CONSUMPTION

Natural gas consumption as a percentage of the total supply of energy (expressed in coal equivalent) was:

1966	1967	1968	1969	1970
9.8	9.3	9.1	10.7	11.5

* NTP - at normal temperature and pressure (i.e. 0 °C and 760 mm. Hg.).

BELGIUM

Principal Oil Statistics

Million metric tons

	1966	1967	1968	1969	1970
Crude and process oils					
Indigenous production	-	-	-	-	-
Imports	16.8	17.2	23.4	28.5	29.9
Total supply	16.8	17.2	23.4	28.5	29.9
Refinery intake	16.6	17.3	23	28.6	30
Exports	0.1	-	-	-	0
Total disposals	16.7	17.3	23	28.6	30
Petroleum products					
Refinery output	15.8	16.3	21.4	26.6	28.2
Output from other sources ...	-	-	-	-	-
Imports	5.8	6.2	5.9	5.5	6.1
Total supply	21.6	22.5	27.3	32.1	34.3
Deliveries to inland consumption	13.7	15.2	17.4	19.6	22.3
Bunkers	2.1	2	2.2	2.9	2.7
Exports	5.3	5.3	7.4	10	9.1
Total disposals	21.1	22.5	27	32.5	34.1
Refinery capacity					
Crude distillation capacity at end year	17.5	25	30.5	34.4	35.9

188

INDIGENOUS PRODUCTION AND IMPORTS OF CRUDE OIL

There is still no indigenous production as all the test drillings carried out to date have been unsuccessful.

Between 1966 and 1970 imports of crude oil have risen considerably as a result of the large-scale increase in refinery capacity reflecting the mounting level of inland consumption and exports.

As from 1968, one inland refinery has been receiving crude through a pipeline from the Belgian coast and a new pipeline recently brought into service connects the Antwerp refineries to the port of Rotterdam.

DEVELOPMENT OF REFINERY CAPACITY

Refinery capacity in Belgium was:

    600,000 t. a year in 1950
  8,750,000 t. a year in 1960
 15,710,000 t. a year in 1965
 30,010,000 t. a year in 1968
 35,885,000 t. a year in 1970

GROWTH IN INLAND CONSUMPTION OF MAIN OIL PRODUCTS

The growth is due to the increase in consumption of three products, namely motor gasoline, and especially, gas/diesel oil and fuel-oil.

The increase in consumption of the latter products is a result of the substitution of oil for coal.

OIL AND NATURAL GAS CONSUMPTION IN RELATION TO TOTAL ENERGY CONSUMPTION

In percentage

FORMS OF ENERGY	1966	1967	1968	1969	1970
Coal	50.15	46.05	42.55	38.01	32.69
Oil	49.31	52.04	53.97	55.78	57.62
Natural gas	0.34	1.45	3.20	6.27	9.31
Hydro and nuclear power	0.20	0.46	0.19	0.06	0.38

BELGIUM

Principal Natural Gas Statistics

Billion[1] cubic metres
at 15°C and 760 mm Hg.

	1966	1967	1968	1969	1970
Indigenous production[2]	–	–	–	–	–
Imports	0.092	0.527	1.406	2.928	4.740
Total supply	0.092	0.527	1.406	2.928	4.740
Inland consumption	0.092	0.527	1.406	2.928	4.740
Exports	–	–	–	–	–
Total disposals	0.092	0.527	1.406	2.928	4.740

1. Billion as used in this Report is 10^9.
2. Excluding waste and flared gas, processing plant fuel, losses and return to formation.

INDIGENOUS PRODUCTION AND IMPORTS OF NATURAL GAS

There is no indigenous production in Belgium.

Natural gas was first imported from the Netherlands in 1966. Since then the volume imported has increased considerably as a result of two factors: firstly the conversion of the distribution network to natural gas, and secondly the increased demand from industry (natural gas is cleaner, it does not pollute the air, and its quality does not vary).

INLAND CONSUMPTION OF NATURAL GAS

The factors underlying the increase in consumption are the same as those applying to imports.

CANADA

Principal Oil Statistics

Million metric tons

	1966	1967	1968	1969	1970
Crude and process oils					
Indigenous production	49	50	54	61	69
Imports	20	23	25	27	29
Total supply	69	73	79	88	98
Refinery intake	49	56	56	58	63
Exports	17	20	23	27	32
Total disposals	66	76	79	85	95
Petroleum products					
Refinery output	46	50	54	56	61
Output from other sources ...	2	2	2	2	3
Imports	8	10	10	10	10
Total supply	56	62	66	68	74
Deliveries to inland consumption	52	56	62	65	68
Bunkers	2	2	3	3	2
Exports	1	2	2	3	4
Total disposals	55	60	67	71	74
Refinery capacity					
Crude distillation capacity at end year	56	59	61	65	70

INDIGENOUS PRODUCTION AND IMPORTS OF CRUDE OIL

Between 1966 and 1970 the average annual rate of growth in the production of crude oil in Canada was about 9 per cent, almost exactly the same rate as in the preceding four-year period. The level reached in 1970 was approximately 69 million metric tons, approximately double the 1961 level.

This strong development of production was due fundamentally to the sustained demand from the domestic refining industry in the central and western part of the country and to the constantly rising demand for Canadian crude in the northern United States. The impact on Canada of disturbances in world oil trade, such as those associated with the 1967 Middle East conflict, was relatively minor and mainly confined to mutually-supporting adjustment of oil trading patterns with the United States.

Exploration for hydrocarbons was largely confined to the Western Provinces in 1966 but in 1970 the most active exploration areas were the Canadian Arctic and the eastern offshore. Important discoveries are anticipated in these so-called "frontier" areas. If these occur, they will inevitably produce deep changes in the Canadian energy economy.

Imports in 1970 of crude oil amounted to 29 million tons and represented 46 per cent of total supply of refinery feedstock. Over the past few years, imports have grown at a slower rate than indigenous oil production, a reflection of the national oil policy which on the one hand has set certain limits to the use of imported petroleum and on the other has encouraged the development of export markets in the United States for oil produced in Western Canada.

DEVELOPMENT OF REFINERY CAPACITY

Recent developments in the oil refining sector include the construction of two large new refineries in the eastern part of the country. In total, 14 million tons/year of new crude oil distillation capacity was installed between end-1966 and end-1970, an increase of 25 per cent over the four-year period. However, a large additional block of capacity has been brought on-stream in the first nine months of 1971, comprising a 2.9 million tons/year plant in Nova Scotia, another of 4.9 million tons/year at Quebec City, and a third of 3.9 million tons/year at Edmonton, Alberta. The last-named refinery replaces a number of older and smaller plants in the Prairie Provinces.

GROWTH IN INLAND CONSUMPTION OF MAIN OIL PRODUCTS

In 1970 the total availability in Canada of refined petroleum products amounted to 74 million metric tons, of which 10 million tons were imported, mainly into the eastern provinces. Products exports and bunkers together accounted for 6 million tons.

Of the 68 million tons of domestic (inland) consumption, the bulk was represented by three main categories of product. Motor gasoline accounted for about 30 per cent of the total, middle distillates including kerosene and diesel oil for another 30 per cent and residual fuel-oils for 23 per cent. Motor gasoline is the fastest-growing major product group of inland consumption, the annual increase in the demand averaging 5.2 per cent between 1966 and 1970. The increase in demand for middle distillates averaged

4.5 per cent annually, and for residual fuel nearly 5 per cent. In the cases of motor gasoline and middle distillates these rates are comparable with those of the preceding four-year period, 1962-1966, but for residual oil the growth rate was little more than half that of the earlier period. This was partly a reflection of the expanded use of natural gas.

OIL CONSUMPTION IN RELATION TO TOTAL ENERGY CONSUMPTION

Between 1966 and 1970 total primary energy consumption (in BTU equivalents) increased at an annual rate of approximately 6.0 per cent, compared with 6.9 per cent in the 1962-1966 period. Petroleum's share in 1962 was 56 per cent, falling to 55 per cent in 1966 and to 53 per cent in 1970. While oil lost some ground to natural gas because of the latter's rapid development in the late 1960s, it remained the dominant source of Canada's energy supply.

193

CANADA

Principal Natural Gas Statistics

Billion[1] cubic metres
at 15°C and 760 mm Hg.

	1966	1967	1968	1969	1970
Indigenous production[2]	31.3	33.8	39.2	45.5	52.4
Imports	1.3	2.0	2.3	1.0	0.3
Total supply	32.6	35.8	41.5	46.5	52.7
Inland consumption	19.9	21.8	24.3	27.1	29.9
Exports	12.2	14.3	16.9	19.3	22.1
Total disposals	32.1	36.1	41.2	46.4	52.0

1. Billion as used in this Report is 10^9.
2. Excluding waste and flared gas, processing plant fuel, losses and return to formation.

INDIGENOUS PRODUCTION AND IMPORTS OF NATURAL GAS

Strong growth in the production of natural gas in Canada continued in the period 1966-1970, and the relative importance of this fuel in the energy economy as a whole further increased. Between 1966 and 1970 the marketable output of natural gas rose by over 21 billion cubic metres, equivalent to a gain over the period of 67 per cent. The reserves supporting this production derive mainly from the gas fields which were discovered during the continuing search for oil and also from the "associated" gas which is or can be produced in conjunction with the exploitation of oil reserves. With the burgeoning demand for gas, especially in the United States, exploration efforts are now being directed more specifically towards discovery of gas.

INLAND CONSUMPTION OF NATURAL GAS

Exports to the United States accounted for over 42 per cent of Canadian natural gas production in 1970, a proportion which has held fairly constant for several years. Thus, the large increase in production reflected a rapid growth in both exports and inland consumption.

Ontario remained the principal domestic gas market area, with over 40 per cent of the net sales being effected in that Province. Extension eastwards of the marketing area is, however, limited by the strong price competition from imported oil fuels. The demand for Canada's natural gas has historically been based on its convenience and price characteristics in competitive uses; it is anticipated that the demand for gas will increasingly be related to the need for relatively pollution-free fuels.

CONSUMPTION OF NATURAL GAS IN RELATION TO TOTAL ENERGY CONSUMPTION

In the period under review, the rate of growth in the inland consumption of natural gas was more than double that for petroleum. Natural gas now supplies approximately 23 per cent of total primary energy consumption in Canada, as against 20 per cent in 1966 and 16 per cent in 1962. Further gains may be anticipated in the residential and commercial sectors of demand and in certain industrial uses.

DENMARK
Principal Oil Statistics

Million metric tons

	1966	1967	1968	1969	1970
Crude and process oils					
Indigenous production	–	–	–	–	–
Imports	5.0	6.6	7.0	9.0	10.2
Total supply	5.0	6.6	7.0	9.0	10.2
Refinery intake	4.8	6.6	6.9	9.0	10.2
Exports	–	–	–	–	–
Total disposals	4.8	6.6	6.9	9.0	10.2
Petroleum products					
Refinery output	4.4	5.9	6.5	8.5	9.7
Output from other sources ...	–	–	–	–	–
Imports	7.9	7.0	8.0	9.1	10.9
Total supply	12.3	12.9	14.5	17.6	20.6
Deliveries to inland consumption	10.7	11.0	12.6	15.4	17.4
Bunkers	0.6	0.7	0.6	0.6	0.5
Exports	0.9	1.1	0.9	1.5	1.9
Total disposals	12.2	12.8	14.1	17.5	19.8
Refinery capacity					
Crude distillation capacity at end year	6.3	6.6	9.2	9.7	10.8

NATURAL GAS

Natural gas is neither produced nor consumed in Denmark.

196

FINLAND

Principal Oil Statistics

Million metric tons

	1966	1967	1968	1969	1970
Crude and process oils					
Indigenous production	-	-	-	-	-
Imports	3.0	5.1	6.0	7.3	9.6
Total supply	3.0	5.1	6.0	7.3	9.6
Refinery intake	4.1	4.7	6.0	7.1	8.2
Exports	-	-	-	-	-
Total disposals	4.1	4.7	6.0	7.1	8.2
Petroleum products					
Refinery output	3.7	4.3	5.5	6.5	7.7
Output from other sources ...	-	-	-	-	-
Imports	3.5	3.0	3.0	2.8	3.0
Total supply	7.2	7.3	8.5	9.3	10.7
Deliveries to inland consumption	6.8	6.9	7.7	9.0	10.0
Bunkers	0.02	0.03	0.03	0.04	0.08
Exports	0.02	0.3	0.2	0.4	0.4
Total disposals	6.8	7.2	7.9	9.4	10.5
Refinery capacity					
Crude distillation capacity at end year	3.0	5.5	5.7	9.0	9.0

INDIGENOUS PRODUCTION AND IMPORTS OF CRUDE OIL

Since there are no oil deposits in Finland, all crude oil has to be imported. The main source in this respect is the Soviet Union, which in 1970 supplied 68 per cent of crude oil imports. In the early 1960's the imports of crude oil accounted for a half of the total oil imports but its share has been increasing steadily so that in 1970 it was 76 per cent.

DEVELOPMENT OF REFINERY CAPACITY

The first refinery in Finland was completed in 1957. Its capacity was about 1 million tons. The second refinery, of a capacity of 2.5 million tons, was completed in 1966. Since then both refineries have been expanded so that to-day the total refinery capacity is 9 million tons. Some plans have been made to increase the refinery capacity in the next few years.

GROWTH IN INLAND CONSUMPTION OF MAIN OIL PRODUCTS

In the 1960's, inland consumption of the main oil products was as follows:

	Million metric tons			Average growth
	1960	1965	1970	
Liquid fuels used in transport sector	0.8	1.5	2.2	10 %
Light fuel-oil	0.5	1.7	3.0	20 %
Heavy fuel-oil	0.9	2.0	4.2	17 %

In 1980 the total inland consumption of oil products is forecast to be 17 to 18 million tons.

OIL CONSUMPTION IN RELATION TO TOTAL ENERGY CONSUMPTION

The share of oil in total energy consumption has increased vigorously. In 1960 it was 22 per cent, in 1965 38 per cent and in 1970 54 per cent. In future, oil consumption will be superseded to some extent by introduction of nuclear power and natural gas. Oil consumption in relation to total energy consumption is forecast to be 53 per cent in 1975 and 57 per cent in 1980.

INDIGENOUS PRODUCTION AND IMPORTS OF NATURAL GAS

There are no natural gas deposits exploitable in Finland, nor has gas so far been imported to the country, either. In 1971 a contract about delivery of natural gas was concluded with the Soviet Union. Import of gas will begin in 1974 and the amounts agreed upon are as follows:

1974   500 million cubic metres per year
1975 1,000 million cubic metres per year
1976 1,100 million cubic metres per year
1977 1,200 million cubic metres per year
1978 1,300 million cubic metres per year
1979 to 1993 1,400 million cubic metres per year

Inland consumption of natural gas

A network for gas delivery will at first be created in south-east Finland and will later be extended to the southern and western parts of the country. Natural gas will be used by industry and power plants. It is proposed to raise the consumption of natural gas in future to 3,000 million cubic metres per year.

Consumption of natural gas in relation to total energy consumption

Consumption of natural gas in relation to total energy consumption will, at least in the beginning, be relatively small, in 1974 to 1980 only 3 to 4 per cent.

Natural gas

Up to now, natural gas is neither produced nor consumed in Finland.

FRANCE

Principal Oil Statistics

Million metric tons

	1966	1967	1968	1969	1970
Crude and process oils					
Indigenous production	2.9	2.8	2.7	2.5	2.3
Imports	64.6	72.8	78.6	88.2	101.3
Total supply	67.5	75.6	81.3	90.7	103.6
Refinery intake	67.5	75.6	81.3	90.7	103.6
Exports	–	–	–	–	–
Total disposals	67.5	75.6	81.3	90.7	103.6
Petroleum products					
Refinery output	62.2	69.7	74.5	84.6	95.6
Output from other sources ...	0.5	0.5	0.5	0.6	0.6
Imports	4.9	4.8	5.1	5.3	6.4
Total supply	67.6	75.0	80.1	90.5	102.6
Deliveries to inland					
consumption	50.0	56.7	63.3	70.9	82.7
Bunkers	2.6	3.1	3.0	4.0	3.9
Exports	13.0	12.6	11.5	11.8	10.7
Total disposals	65.6	72.4	77.8	86.7	97.3
Refinery capacity					
Crude distillation capacity at end year	79.2	83.8	97.3	105.2	116.5

200

Indigenous production of crude oil

At 31st December, 1970, accrued production of crude oil from indigenous resources amounted to 39 million tons.

On that date recoverable reserves in these fields were estimated at: crude oil: 12.5 million tons, and natural gas liquids: 13 million tons, — i.e. a total of 25.5 million tons of liquid hydrocarbons.

At the end of 1971 the average rate of oil production from indigenous fields was 1.7 million tons a year.

The largest fields are in the Aquitaine area.

Production from the fields now being exploited is expected to fall progressively in the future and will probably amount to about 1 million tons a year in 1975.

Prospects of reconstituting indigenous reserves are poor insofar as possible onshore fields are concerned. Offshore, i.e., on the Continental Shelf in the Atlantic and the Mediterranean, the outlook is more encouraging.

Imports of crude oil

As a result of the considerable increase in the demand for oil products from 1966 to 1970 and the steady decrease in the small-scale indigenous production of liquid hydrocarbons, the relative share of the latter in the country's total oil supplies fell from 4.3 per cent in 1966 to 2.3 per cent in 1970. The volume of imports has accordingly risen considerably over this period (by about 60 per cent).

This increase is likely to continue in the next few years. For 1975, imports are forecast at 145/160 million tons.

Between 1966 and 1970 France drew its supplies mainly from the Western Mediterranean (over 40 per cent from Algeria and Libya), the Eastern Mediterranean (about 15 per cent, from Iraq and Saudi Arabia) and the Persian Gulf (about 28 per cent).

Over the same period imports from Algeria and the Eastern Mediterranean have steadily decreased whilst deliveries from Libya and the Persian Gulf have increased.

In the next few years it is expected that the proportion of Algerian crude in French supplies will decrease appreciably and that imports from the Persian Gulf will increase, to reach about 50 per cent of total imports in 1975. Imports from the Eastern Mediterranean should remain at the same level as in 1970 (about 12 per cent). Imports from Africa south of the Sahara will probably be stepped up and by 1975 they could well account for about 12 per cent of the country's supplies.

Development of refinery capacity

The average yearly growth rate in refinery capacity from 1966 to 1970 was 11 per cent.

Half of this increase in capacity was due to the expansion of existing refineries and the other half to five new refineries being built at inland sites, each with a capacity of about 3.6 million tons a year.

This has had little effect on the geographical distribution of refinery capacity which is mainly located in the Paris area, the North, the East, the Rhone Basin, on the Mediterranean coast, and in the West and South-West.

In the next few years refinery capacity is expected to increase on average by some 9 per cent a year, mainly as a result of the expansion of existing refineries (75 per cent), rather than the construction of new ones (25 per cent).

The average size of new refineries will probably increase to some 6 million tons a year.

GROWTH IN INLAND CONSUMPTION OF MAIN OIL PRODUCTS

The following table shows the growth in consumption of the main oil products in France from 1966 to 1970, together with forecasts for 1975:

Million metric

	1966	1967	1968	1969	1970	1975 FORECAS'
Motor gasoline	9.1	9.9	10.7	11.3	12.3	17.1
Diesel oil	3.2	3.5	3.9	4.3	4.7	7.3
Gas oil	16.0	19.6	23.1	25.9	29.6	43.5
Residual fuel oil	11.2	12.8	13.6	16.2	21.0	38.7

The steady growth in the consumption of motor gasoline and gas/diesel oil from 1966 to 1970 is expected to continue but at a slightly lower rate between 1970 and 1975. The increase in the consumption of gas-oil and residual fuel-oil was particularly large from 1966 to 1970 as these products have gradually replaced coal for domestic heating and industrial purposes. This development and its repercussions on the consumption of other forms of energy are tapering off, so that the rate of increase in fuel-oil consumption should tend to come into line with the normal increase in requirements for industrial purposes and domestic heating.

Diesel oil and gas-oil together account for 45 per cent of the total consumption of oil products; residual fuel-oil accounts for about 25 per cent and motor gasoline 15 per cent, but the latter is tending to diminish slightly.

The following is a comparative table of growth rates for the main oil products in France from 1966 to 1970 and includes forecasts for the period 1970 to 1975.

In percentage

	MOTOR GASOLINE	DIESEL OIL	GAS OIL (DOMESTIC FUEL OIL)	RESIDUAL FUEL OIL
Average annual growth rate 1966/1970	8	10	16.5	17
Average annual growth rate 1970/1975	7	9	8	13

The next table shows the relative importance of oil products in consumption of all forms of energy in France from 1966 to 1970, together with a forecast of the breakdown of the main forms of energy for 1975.

In percentage

	1966	1967	1968	1969	1970	1975
Solid fuels	37.2	35.2	32.3	30	25.6	13.5
Natural Gas	4.9	5.1	5.7	6.1	6.4	8.5
Oil products used for energy purposes ...	47.3	50.6	52.5	54.8	58.9	69
Primary electricity	10.6	9.1	9.5	9.1	9.1	9
Total consumption of primary energy:						
- %	100	100	100	100	100	100
- mtce	174.1	183	193	207.5	224.5	280

The rundown of coal from 1966 to 1970 would appear to have been offset mainly by the increase in the proportion of oil products which has risen from 47 to 59 per cent of total energy consumption during the period under review.

In view of the disparity between coal and fuel-oil prices, the conversion of many industrial plants from coal to fuel-oil is likely to continue up to 1975.

With the continuing growth foreseen in the consumption of gasoline and gas-diesel oil, the share of oil products in total energy consumption will increase considerably up to 1975 despite the growing market for natural gas. By 1975 oil products could well account for two-thirds of total energy consumption.

The comparative table below shows rates of increase in the consumption of energy products in France from 1966 to 1970, with forecasts for the period 1970-1975.

In percentage

	SOLID FUELS	NATURAL GAS	OIL PRODUCTS USED FOR ENERGY PURPOSES	PRIMARY ELECTRICITY	TOTAL CONSUMPTION OF PRIMARY ENERGY
Average annual growth rate 1966/1970	-2.9	+14	+12.5	+2.3	+6.6
Forecast average annual growth rate 1970/1975	-8	+11	+8	+5	+4.5

FRANCE

Principal Natural Gas Statistics

Billion[1] cubic metres
at 15°C and 760 mm Hg.

	1966	1967	1968	1969	1970
Indigenous production[2]	5.2	5.6	5.7	6.6	6.9
Imports	0.5	0.9	1.9	2.8	3.8
Total supply	5.7	6.5	7.6	9.4	10.7
Inland consumption	4.7	5.3	6.2	7.4	11.9
Exports	-	-	-	-	-
Total disposals	4.7	5.3	6.2	7.4	11.9

1. Billion as used in this Report is 10^9.
2. Excluding waste and flared gas, processing plant fuel, losses and return to formation.

INDIGENOUS PRODUCTION OF NATURAL GAS

At 31st December, 1970 total cumulative production of natural gas amounted to about 90 billion cubic metres of marketable gas.

At the same date, proven reserves of natural gas were estimated at about 200 billion cubic metres.

At the end of 1971 the average annual production was about 11 billion cubic metres of marketable gas. The fields are located in Aquitaine, the Pau region (Lacq and Meillon) and the Boussens region (St. Marcet and Auzan).

Production is expected to rise by about 10 per cent in 1972 and level out by 1974-1975 and then fall off slightly as the St. Marcet field becomes depleted. The largest fields (Lacq and Meillon), however, should show no change up to 1980.

Exploration in new areas and in particular in the Pyrénées region may well lead to further gas finds.

IMPORTS OF NATURAL GAS

In 1970 France imported 3.5 billion cubic metres of "normal" natural gas from the Netherlands (3 billion m³) and Algeria (0.5 billion m³).

Imports are likely to increase substantially and may well reach 11.5 billion cubic metres in 1975 and 14.6 billion cubic metres in 1980.

In addition to the rising scale of deliveries under current contracts with the Netherlands, additional supplies are expected from Algeria from 1972 onwards (3.5 billion a year) and the USSR from 1977 onwards (2.5 billion m³ a year).

INLAND CONSUMPTION OF NATURAL GAS

Consumption of natural gas has increased subtantially in France over the last five years.

This growth has been partly the result of the price differential between natural gas and fuel-oil, although this is much less than between fuel-oil and coal.

Natural gas consumption is likely to continue to increase in the next few years.

CONSUMPTION OF NATURAL GAS IN RELATION TO TOTAL ENERGY REQUIREMENTS

The share of natural gas in total energy consumption in France is to be seen in the table given in the preceding commentary on the oil statistics.

The growth of natural gas consumption, which was very high over the period 1966-1970, will continue in the next few years (cf. comparative table of growth rates of the various forms of energy given in the preceding commentary on oil statistics), as a result of the general increase in energy consumption and the sustained demand for this form of energy.

FEDERAL REPUBLIC OF GERMANY

Principal Oil Statistics

Million metric tons

	1966	1967	1968	1969	1970
Crude and process oils					
Indigenous production	7.9	7.9	8.0	7.9	7.5
Imports	67.7	72.0	84.1	89.6	98.8
Total supply	75.6	79.9	92.1	97.5	106.3
Refinery intake	75.3	79.6	90.8	97.4	107.1
Exports	-	-	-	-	-
Total disposals	75.3	79.6	90.8	97.4	107.1
Petroleum products					
Refinery output	69.8	73.7	83.7	89.7	98.5
Output from other sources ...	0.7	0.5	0.6	0.6	0.5
Imports[1]	18.5	19.3	20.6	25.1	30.4
Total supply	89.0	93.5	104.9	115.4	129.4
Deliveries to inland consumption[2]	79.7	82.7	92.1	103.7	116.2
Bunkers	3.9	3.4	3.7	4.1	3.8
Exports	5.5	6.4	7.8	7.5	8.4
Total disposals	89.1	92.5	103.6	115.3	128.4
Refinery Capacity					
Crude distillation capacity at end year	88.7	109.3	113.1	115.1	120.3

1. Including supplies from the German Democratic Republic.
2. Including refineries' own consumption and military consumption.

INDIGENOUS PRODUCTION AND IMPORTS OF CRUDE OIL

Production of crude oil in the Federal Republic of Germany is stagnant with a declining tendency. In the period under review, from 1966 to 1970, it reached its record level of 7.982 million metric tons in 1968. However, at that time its share in total crude oil consumption in the Federal Republic had already declined to 8.7 per cent after the 1966 share of 10.4 per cent and amounted to no more than 7.1 per cent in 1970.

In imports a certain shift from Persian Gulf crudes to crudes of African origin was to be seen. The share of African crudes in total crude oil imports came to 58.8 per cent (1966: 50.9 per cent) and that of Middle Eastern crudes to 34.2 per cent (1966: 38.8 per cent). The reason was above all the extraordinary increase in imports from Libya between 1966 and 1968. However, imports from Lybia have reached their peak. The proportion of Libyan oil in total imports is presently showing a declining tendency (1970: 41.4 per cent).

DEVELOPMENT OF REFINERY CAPACITY

In the years from 1966 to 1970, the refinery capacity in the Federal Republic rose by 35.6 per cent to 120 million tons, but did not keep pace with the rise of consumption of oil products in that same period. Therefore, demand had to be satisfied by increased imports of finished products. In 1970 imports rose by 64.2 per cent to 30.4 million tons, which corresponds to a share of imports in total domestic sales (including bunkers) from 21 per cent (1966) to 23.9 per cent (1970).

The development towards a decentralization of refinery locations and the establishment of capacities close to the consumption centres is progressing further. While in 1966 the main refinery centres were still found in the Rhine/Ruhr area (39.8 per cent) and Northern Germany (30.0 per cent), they also spread to other areas in the federal territory in 1970: of the present refinery capacities, 32.9 per cent are located in the Rhine/Ruhr area, 26.5 per cent in the Rhine/Main area and the South West of the country, 22.4 per cent in Northern and 18.2 per cent in Southern Germany.

These locations close to the centres of consumption are ultimately a consequence of the extension of pipeline capacities by which 81 per cent of all crude oil imports were transported to the refineries in 1970. Two major long-distance crude oil pipelines were put into operation since 1964 in addition to several branch mains: the pipeline Genoa-Ingolstadt (CEL) (1966) and the pipeline Trieste-Ingolstadt (TAL) (1967). Both are supplying the refinery centre in Ingolstadt and— after the reversal of the flow direction of the Rhine-Danube pipeline— the Karlsruhe refinery centre.

GROWTH IN INLAND CONSUMPTION OF MAIN OIL PRODUCTS

In the period from 1966 to 1970, domestic consumption of oil products, including military consumption and refineries' own consumption as well as bunkers, increased by 44.5 per cent, amounting to 127.5 million tons in 1970. A major share of the increase was supplied by light fuel-oil— which rose by 64.2 per cent— with consumption (including military, refineries' own consumption and bunkers rising from 26.9 million tons (30.4 per cent) to 44.1 million tons (34.6 per cent). The combined share of light and heavy fuel-oils increased from 53.0 million tons (60.0 per cent) to 78.0 million tons (61.2 per cent). On the other hand the share of gasoline declined further from 14.9 per cent to 13.5 per cent and that of diesel fuel from 10.0 per cent to 8.4 per cent in the same period.

OIL CONSUMPTION IN RELATION TO TOTAL ENERGY CONSUMPTION

While total consumption of primary energy rose by 26.4 per cent from 1966 to 1970, an increase in oil consumption (excluding bunkers) of 46.5 per cent was registered. In this period, about four fifths of the total increase in primary energy consumption was accounted for by oil.

As a result of its higher relative growth, oil raised its share in total energy supply in the Federal Republic of Germany to 53.0 per cent (1966: 45.7 per cent).

This development shows clearly the favourable competitive situation of oil in the energy market. However, as the dependence on oil increases, the problem of the security of supplies is intensified, thus becoming a principal consideration.

GERMANY

Principal Natural Gas Statistics

Billion[1] cubic metres
at 15°C and 760 mm Hg.

	1966	1967	1968	1969	1970
Indigenous production[2]	3.18	4.03	6.13	8.58	12.32
Imports	0.05	0.34	1.54	2.67	3.74
Total supply	3.23	4.37	7.67	11.25	16.06
Inland consumption	3.23	4.37	7.67	11.25	16.05
Exports	-	-	-	-	0.01
Total disposals	3.23	4.37	7.67	11.25	16.06

1. Billion as used in this Report is 10^9.
2. Excluding waste and flared gas, processing plant fuel, losses and return to formation.

INDIGENOUS PRODUCTION AND IMPORTS OF NATURAL GAS

In recent years, production of natural gas increased extraordinarily in the Federal Republic of Germany. The average rate of growth rose to 45 per cent in the last three years. This result is explained by the large-scale investments for exploration and additional exploitation drillings. Another vigorous increase in production may be expected till the mid-seventies.

Imports of natural gas have also increased sharply. As compared to indigenous production, these played a minor role in the years 1966 and 1967. However, in 1968 imports secured a share of 20 per cent, registering an increase of 24 per cent in both 1969 and in 1970. By 1975, imports from the Netherlands and the USSR are likely to have reached the level of indigenous production.

INLAND CONSUMPTION OF NATURAL GAS

There is hardly any economic branch nowadays which does not use natural gas. The largest consumption was recorded for the iron making and the chemical industries. In the latter, natural gas is gaining importance as a base material. The use of natural gas for electricity production is also on the increase. In the household, commerce and small industry areas the growth rates are also substantial. Increasing sales of natural gas have also to be expected as a result of the conversion from town to natural gas. Consumption by households in particular is likely to rise as a consequence of the conversion to natural gas.

CONSUMPTION OF NATURAL GAS IN RELATION TO TOTAL ENERGY CONSUMPTION

In 1966, the share of natural gas in total primary energy consumption amounted to 1.2 per cent. The corresponding share registered in 1970 rose to 5.4 per cent.

GREECE

Principal Oil Statistics

Million metric tons

	1966	1967	1968	1969	1970
Crude and process oils					
Indigenous production	-	-	-	-	-
Imports	3.1	3.9	4.4	4.5	5.0
Total supply	3.1	3.9	4.4	4.5	5.0
Refinery intake	3.0	4.0	4.3	4.5	5.0
Exports	-	-	-	-	-
Total disposals	3.0	4.0	4.3	4.5	5.0
Petroleum products					
Refinery output	2.8	3.6	4.0	4.2	4.7
Output from other sources ...	-	-	-	-	-
Imports	1.9	1.6	1.5	1.8	1.8
Total supply	4.7	5.2	5.5	6.0	6.5
Deliveries to inland consumption	3.7	4.3	4.8	5.2	5.5
Bunkers	0.9	0.9	0.6	0.6	0.7
Exports	-	0.2	0.2	0.2	0.2
Total disposals	4.6	5.4	5.6	6.0	6.4
Refinery capacity					
Crude distillation capacity at end year	4.4	4.4	4.4	4.4	5.1

211

Indigenous production and imports of crude oil

Up to now there is no indigenous production of crude oil in Greece. Since 1967 several contracts have been signed between the Greek State and private enterprises for oil exploration in Greece. Off-shore drilling is already under way in several places in the Northern Aegean Sea.

The Country's crude requirements are covered partially by a long-term supply contract between the Greek State and the major international oil companies. With the Suez Canal closed all the above mentioned crude comes from Eastern Mediterranean ports (Kirkuk and Arabian Light). The Esso Refinery in Salonica is also processing its own crude of the same origin as above. The balance of the country's requirements (15-20 per cent) is covered by Soviet crude (Tuimaza) purchased by the State within the framework of a Clearing Agreement of Greece with the USSR.

Development of refinery capacity

1958 — Start-up of the country's first State-owned refinery in Aspropyrgos, Attica. This refinery can process approximately 1,900,000 metric tons of crude per year.

1966 — Second refinery owned by Esso-Pappas started operation early in Spring. Crude processing capacity over 2,500,000 metric tons per year.

1970 — Esso-Pappas Salonica Refinery increases capacity by debottlenecking to 3,200,000 metric tons per year. Plans of Aspropyrgos for a major expansion and modernization under way. Expanded refinery to be owned 2/3 by Niarchos Group and 1/3 by Greek State. Planning completed for the erection of two export refineries in Attika area.

Growth in inland consumption of main oil products

	PERCENTAGE INCREASE (DECREASE)			
	1966-67	1967-68	1968-69	1969-70
Products				
1. Total	16.0	10.6	9.2	3.9
2. LPG	27.8	15.3	10.2	9.3
3. Motor Gasoline	8.9	15.9	9.2	11.5
4. Aviation Fuels	21.7	11.9	19.3	14.2
5. Kerosene	2.2	(9.0)	(9.4)	(8.0)
6. Diesel	15.0	7.5	9.3	6.1
7. Fuel Oil	18.2	12.5	7.8	(1.0)*

* Fuel oil increase is greatly influenced by fluctuations of Public Power Corporation demand on account of Hydro-power availabilities per year.

The contribution of oil products in covering the final demand of energy in Greece is in the order of 79 per cent for the year 1970.

NATURAL GAS

Natural gas in neither produced nor consumed in Greece.

213

ICELAND
Principal Oil Statistics

Million metric tons

	1966	1967	1968	1969	1970
Crude and process oils					
Indigenous production					
Imports					
Total supply	-	-	-	-	-
Refinery intake					
Exports					
Total disposals	-	-	-	-	-
Petroleum products					
Refinery output	-	-	-	-	-
Output from other sources ...	-	-	-	-	-
Imports	0.5	0.5	0.6	0.4	0.5
Total supply	0.5	0.5	0.6	0.4	0.5
Deliveries to inland consumption	0.5	0.5	0.6	0.4	0.5
Bunkers	-	-	-	-	-
Exports	-	-	-	-	-
Total disposals	0.5	0.5	0.6	0.4	0.5
Refinery capacity					
Crude distillation capacity at end year	-	-	-	-	-

NATURAL GAS

Natural gas is neither produced nor consumed in Iceland.

IRELAND

Principal Oil Statistics

Million metric tons

	1966	1967	1968	1969	1970
Crude and process oils					
Indigenous production	-	-	-	-	-
Imports	1.6	2.6	2.3	2.2	2.7
Total supply	1.6	2.6	2.3	2.2	2.7
Refinery intake	1.6	2.6	2.3	2.3	2.7
Exports	-	-	-	-	-
Total disposals	1.6	2.6	2.3	2.3	2.7
Petroleum products					
Refinery output	1.5	2.5	2.2	2.2	2.6
Output from other sources ...	-	-	-	-	-
Imports	1.2	1.3	1.4	2.0	2.3
Total supply	2.7	3.8	3.6	4.2	4.9
Deliveries to inland consumption	2.4	2.8	2.9	3.3	3.8
Bunkers	0.1	0.2	0.1	0.1	0.2
Exports	0.2	0.8	0.6	0.7	0.9
Total disposals	2.7	3.8	3.6	4.1	4.9
Refinery capacity					
Crude distillation capacity at end year	2.5	2.5	2.5	2.5	2.5

215

Indigenous production and imports of crude oil

Prospecting for oil was undertaken in 1962/63. Six holes were drilled in widely scattered areas throughout the country but there were no tangible results. Exploration is at present being carried out in this country's share of the Continental Shelf outside territorial waters. Drilling commenced on a well, 28 miles off Kinsale on 1st May, 1970. Traces of hydrocarbons were found in non-commercial quantities. The well was completed at 10,500 ft. plugged and abandoned. A further well is being drilled at present off the Waterford coast. It is expected that fourteen licences for general surveying of the country's designated shelf area, excluding areas already held under exclusive exploration licence, will be granted shortly.

Imports of crude oil are mostly from the Middle East and have increased from 1,616,000 t in 1966 to 2,722,000 t in 1970, an increase of 68 per cent.

Development of refinery capacity

There is one refinery in Ireland, the annual capacity of which was raised by 0.5 million tons in 1966 to a total of 2.5 million tons. The growth in consumption of petroleum products in Ireland suggests that there may be scope for a further increase in refining capacity within the country.

Growth in inland consumption of main oil products

The upward trend in consumption of oil products in this country continues. Inland consumption in 1970 at 3,819,000 metric tons was up by 56 per cent on 1966. The increase in the consumption of the three main products are as follows:

- a) Motor Gasoline from 433,000 metric tons in 1966 to 617,000 metric tons in 1970;
- b) Gas/Diesel Oil from 443,000 metric tons in 1966 to 776,000 metric tons in 1970; and
- c) Fuel-Oil from 1,232,000 metric tons in 1966 to 1,913,000 metric tons in 1970.

Present trends suggest that the consumption of these main products will increase as follows:

- — Motor Gasoline to 934,000 metric tons by 1975;
- — Gas/Diesel Oil to 1,253,000 metric tons by 1975 and;
- — Fuel-Oil to 2,944,000 metric tons by 1975.

Oil consumption in relation to total energy consumption

In 1970, oil accounted for about 61 per cent of the country's total requirements of primary energy, compared with 48 per cent in 1966.

Natural gas

Natural gas is neither produced nor consumed in Ireland.

ITALY

Principal Oil Statistics

Million metric tons

	1966	1967	1968	1969	1970
Crude and process oils					
Indigenous production	1.8	1.6	1.5	1.5	1.4
Imports	79.2	84.3	92.6	102.5	114.0
Total supply	81.0	85.9	94.1	104.0	115.4
Refinery intake	80.6	85.8	94.1	103.6	115.2
Exports	-	-	-	-	-
Total disposals	80.6	85.8	94.1	103.6	115.2
Petroleum products					
Refinery output	76.4	81.4	89.1	99.0	111.6
Output from other sources ...	-	-	0.7	0.8	0.8
Imports	3.0	3.5	4.0	3.9	3.4
Total supply	79.4	84.9	93.8	103.7	115.8
Deliveries to inland consumption	45.5	50.8	59.9	66.4	76.2
Bunkers	8.2	7.8	7.9	8.6	7.9
Exports	21.1	22.6	24.8	27.6	28.8
Total disposals	74.8	81.2	92.6	102.6	112.9
Refinery capacity					
Crude distillation capacity at end year[1]	83.5	88.5	101.5	104.4	104.4

1. Capacity authorised by law. Does not include petrochemical plants.

Indigenous production and imports of crude oil

Indigenous crude oil production is continuing at a steady level. It is on a modest scale, and makes only a small contribution to the country's oil needs.

In recent years crude oil imports have risen by about 10 per cent a year, due mainly to the increase in inland consumption.

Development of refinery capacity

Refinery capacity is matched to the demand for oil products from Italian refineries to cover inland consumption and to meet all export requirements.

It will likewise have to keep pace with the above needs in the future.

Growth in inland consumption of main oil products

In recent years inland consumption viewed as a whole has risen by 10 per cent a year on average.

This rate of increase is expected to be maintained in the next few years unless there is, for example, a slowdown in industrial growth or a need to diversify the forms of energy used for reasons of security of supplies.

Oil consumption in relation to total energy consumption

Oil now accounts for about 70 per cent of total energy consumption, compared with about 60 per cent in 1968.

Again, subject to the reservations made under the previous heading, a similar rate of increase is to be expected over the next few years.

ITALY

Principal Natural Gas Statistics

Billion[1] cubic metres
at 15°C and 760 mm Hg.

	1966	1967	1968	1969	1970
Indigenous production[2]	8.8	9.4	10.4	11.7	13.1
Imports	–	–	–	–	–
Total supply	8.8	9.4	10.4	11.7	13.1
Inland consumption	8.4	9.2	10.7	11.8	12.9
Exports	–	–	–	–	–
Total disposals	8.4	9.2	10.7	11.8	12.9

1. Billion as used in this Report is 10^9.
2. Excluding waste and flared gas, processing plant fuel, losses and return to formation.

INDIGENOUS PRODUCTION AND IMPORTS OF NATURAL GAS

National gas production remains substantially unchanged although a slight upward trend is discernable.

Imports of natural gas are expected to rise steeply. Those from Libya will come to some 3,000 million cubic metres a year, over a period of 20 years. From 1973 on, it also planned to import gas at a rate of up to some 6,000 million cubic metres a year over a 20-year period from both Russia and the Netherlands respectively.

INLAND CONSUMPTION OF NATURAL GAS

Inland consumption of natural gas is likely to rise substantially, as and when greater quantities of imported natural gas become available.

CONSUMPTION OF NATURAL GAS IN RELATION TO TOTAL ENERGY CONSUMPTION

Natural gas accounts for about 8 to 9 per cent of total energy consumption.

This percentage will no doubt increase over the next few years, in line with the growing imports of natural gas, as mentioned earlier.

JAPAN

Principal Oil Statistics

Million metric tons

	1966	1967	1968	1969	1970
Crude and process oils					
Indigenous production	0.7	0.7	0.7	0.7	0.8
Imports	97.7	103.3	121.5	144.1	169.5
Total supply	98.4	104.0	122.2	144.8	170.3
Refinery intake	84.2	102.6	115.6	137.6	169.2
Exports	1.6	–	–	–	–
Total disposals	85.8	102.6	115.6	137.6	169.2
Petroleum products					
Refinery output	82.1	101.1	114.1	139.7	159.7
Output from other sources ...	1.5	1.7	2.2	9.8	1.0
Imports	13.2	14.8	17.7	18.6	27.4
Total supply	96.8	117.6	134.0	168.1	188.1
Deliveries to inland consumption	83.3	102.6	120.1	143.3	176.2
Bunkers	11.3	9.7	7.4	9.6	6.7
Exports	1.6	2.5	2.4	12.1	9.5
Total disposals	96.2	114.8	129.9	165.0	192.4
Refinery capacity					
Crude distillation capacity at end year[1]	2.2	2.3	2.8	3.2	3.7

1. In million barrels per stream day.

INDIGENOUS PRODUCTION AND IMPORTS OF CRUDE OIL

The level of the crude oil production in Japan has been about 0.7 million tons per year for these several years and it is not expected to be raised even in the future since there has been no discovery of new oil fields. However, the exploration for oil is now actively proceeding on the continental shelves surrounding Japan, the results of which are yet to be seen.

Therefore, the oil demand in Japan is almost completely supplied by the imports of crude oil and the percentage of imported crude oil against the total supply was 99.0 per cent in 1966 and 99.6 per cent in 1970. During the period from 1966-1970, the volume of imported crude oil almost doubled, which corresponds to the growth rate of 18.4 per cent per year.

The main source for crude oil supply to Japan is the Middle East area and the percentage of crude oil supply from this area was 89.4 per cent in 1966 and 85.5 per cent in 1970. The slight decrease in the percentage of the Middle East in recent years reflects the increased imports of low-sulphur crude oil from South East Asia centering around Indonesia due to the intensified necessity for the prevention of public nuisance in Japan. It is noted in this respect that the percentage of crude oil imports from the South East Asia was raised from 6.2 per cent in 1966 to 13.3 per cent in 1970.

DEVELOPMENT OF REFINERY CAPACITY

Since the larger portion (about 85 per cent) of the domestic demand for oil products is being supplied through refining in Japan, her refining capacity is being raised in parallel with the increase in the oil demand. The average growth rate of the Japan's refining capacity was 13.5 per cent during the 1966-1970 period. Furthermore, in view of the economical merit to be obtained from using large-sized ocean tankers, taking added value of oil by means of refining or other reasons, refining for Japan's oil demand is expected to continue to be made in her territory in the future and therefore Japan's refining capacity will continue to increase in line with the growth of the oil demand.

From the viewpoint of the prevention of air pollution, the desulfurizing capacity has been increased year by year and its capacity was increased from 40,000 BSD in 1967 for hydro-desulfurization of residual oil to 113,000 BSD and 156,000 BSD for hydro-desulfurization of residual oil and for that of gas-oil respectively in 1970.

GROWTH IN INLAND CONSUMPTION OF MAIN OIL PRODUCTS

Japan's domestic consumption of oil products doubled during the 1966-1970 period with an annual growth rate of about 20 per cent. If each product is considered, the highest growth rate of 29.9 per cent (average annual growth rate during the 1966-1970 period) is shown by naphtha for petrochemical use reflecting the development of its user industries including synthetic textiles, electric equipment for home use, automobiles, etc. Kerosene also showed a remarkable 26.5 per cent annual growth in the same period with the increase in its demand for home heating purpose in recent years, while fuel-oil showed 20.6 per cent reflecting the economic activity

221

as a whole.   As to the other main products, gasoline and gas-oil showed an annual growth rate of 14.4 per cent and 17.0 per cent respectively.

OIL CONSUMPTION IN RELATION TO TOTAL ENERGY CONSUMPTION

Total energy consumption increased at an annual rate of 14.2 per cent during the 1966-1970 period.   However, since the growth rate of coal and hydraulic power generation only showed 7.6 per cent and 0.1 per cent respectively, the percentage of oil against total energy consumption increased from 60.4 per cent in 1966 to 70.8 per cent in 1970.

In the future, the growth rate of coal is expected to decrease while that of hydraulic power generation is anticipated to stay at its present level. Further, it is now estimated that atomic power generation will not develop so rapidly in the coming ten years and the increased demand for energy in the future will inevitably have to depend upon the supply of oil.

JAPAN

Principal Natural Gas Statistics

Billion[1] cubic metres
at 15°C and 760 mm Hg.

	1966	1967	1968	1969	1970
Indigenous production[2]	1.8	1.9	2.1	2.3	2.4
Imports	-	-	-	0.3	1.4
Total supply	1.8	1.9	2.1	2.6	3.8
Inland consumption	1.8	1.9	2.1	2.6	3.8
Exports	-	-	-	-	-
Total disposals	1.8	1.9	2.1	2.6	3.8

1. Billion as used in this Report is 10^9.
2. Excluding waste and flared gas, processing plant fuel, losses and return to formation.

INDIGENOUS PRODUCTION AND IMPORTS OF NATURAL GAS

In Japan natural gas is being produced in the areas of Akita, Yamagata, Niigata, etc. on the Japan Sea coast and in the Chiba area in the Kanto District. At present, the areas that are producing natural gas are rather limited but as the exploration and development of natural gas on the continental shelves around Japan are being actively pursued, the natural gas producing area is expected to expand from inland to the sea area hopefully resulting in an increase in the volume produced.

Since the liquefying technique for natural gas has been developed and the ocean transport of natural gas by LNG tankers commercially realized, Japan started to import LNG from Alaska in the latter half of 1969 unloading it at Negishi, Yokohama. Japan also has plans to import LNG from Brunei and other countries.

INLAND CONSUMPTION OF NATURAL GAS

The Japan's total production of natural gas in 1970 amounted to about 2,400 million m^3 and this is used as the raw material for the chemical industry and for city gas manufacturing in the Japan Sea coast area near the gas fields and in the Kanto area. After the imports of LNG were started, the volume of natural gas to be used for power generation and city gas manufacturing has been increased.

223

CONSUMPTION OF NATURAL GAS IN RELATION TO TOTAL ENERGY CONSUMPTION

Imports of LNG to Japan were started in November 1969 and the imported volume of LNG reached 260 million m³ and 1,400 million m³ in 1969 and 1970 respectively. At present, this volume is mainly supplied for power generation and city gas manufacturing. Namely, of the imported volume of 1,400 million m³ in 1970, about 75 per cent (about 970 million m³) was consumed for power generation and about 25 per cent (about 330 million m³) was consumed for city gas manufacturing.

The percentage of natural gas against the total energy consumption is now about 1.3 per cent.

LUXEMBOURG

Principal Oil Statistics

Million metric tons

	1966	1967	1968	1969	1970
Crude and process oils					
Indigenous production	–	–	–	–	–
Imports	–	–	–	–	–
Total supply	–	–	–	–	–
Refinery intake	–	–	–	–	–
Exports	–	–	–	–	–
Total disposals	–	–	–	–	–
Petroleum products					
Refinery output	–	–	–	–	–
Output from other sources ...	–	–	–	–	–
Imports	0.96	1.06	1.20	1.30	1.38
Total supply	0.96	1.06	1.20	1.30	1.38
Deliveries to inland consumption	0.95	1.05	1.15	1.28	1.36
Bunkers	–	–	–	–	–
Exports	–	–	–	–	–
Total disposals	0.95	1.05	1.15	1.28	1.36
Refinery capacity					
Crude distillation capacity at end year	–	–	–	–	–

From 1966 to 1970 overall consumption of oil products thus increased by 0.4 million tons, i.e. 43 per cent.

With regard to the main products consumed, the following increases were recorded from 1966 to 1970:

Motor gasoline .... + 25.6 per cent
Gas/diesel-oil .... + 33.6 per cent
Light fuel-oil .... + 88.9 per cent
Residual fuel-oil .... + 36.5 per cent

The increased consumption of gas/diesel oil and light fuel-oil is due to the more general use of these fuels for domestic heating and that of residual fuel-oil to growing industrial requirements, especially in the iron and steel industry.

The proportion of the total overall consumption of primary and equivalent energy sources accounted for by oil products was 24.7 per cent in 1966 and about 29 per cent in 1970.

NATURAL GAS

As Luxembourg will not be connected to the natural gas network until 1972, there is neither production nor consumption of natural gas for the time being.

THE NETHERLANDS

Principal Oil Statistics

Million metric tons

	1966	1967	1968	1969	1970
Crude and process oils					
Indigenous production	2.4	2.3	2.1	2.0	1.9
Imports	31.7	33.8	38.2	50.8	60.1
Total supply	34.1	36.1	40.3	52.8	62.0
Refinery intake	34.0	35.2	39.6	51.9	62.2
Exports	–	–	0.1	0.5	–
Total disposals	34.0	35.2	39.7	52.4	62.2
Petroleum products					
Refinery output	31.6	32.9	36.5	48.6	58.5
Output from other sources ...	–	–	–	–	–
Imports	11.1	9.4	7.9	6.2	9.6
Total supply	42.7	42.3	44.4	54.8	68.1
Deliveries to inland co consumption	19.0	19.9	20.8	21.8	24.0
Bunkers	5.0	5.2	6.6	8.2	8.8
Exports	17.9	17.3	16.5	24.6	33.7
Total disposals	41.9	42.4	43.9	54.6	66.5
Refinery capacity					
Crude distillation capacity at end year	33.2	37.8	42.2	67.2	68.5

227

INDIGENOUS PRODUCTION AND IMPORTS OF CRUDE OIL

Indigenous production is declining slowly.

Imports of crude oil roughly doubled in the five-year period under consideration. This fast growth is associated with the expansion of the refinery capacity, mainly for export. In 1966 indigenous production accounted for 12.5 per cent of inland consumption, compared with 8 per cent in 1970.

DEVELOPMENT OF REFINERY CAPACITY

Refinery capacity more than doubled, from 33.2 million tons in 1966 to 68.5 million tons in 1970. This was mainly a result of expansion schemes at existing refineries.

The Shell refinery was increased from 17 to 24.5 million tons, the Esso refinery from 7.5 to 16 million tons, the Chevron refinery from 5 to 12.9 million tons and the Gulf refinery from 3.5 to 5 million tons. BP and Mobil have built new refineries with capacities of 5.8 and 4 million tons, respectively, at the end of 1970. After completion of the schemes announced to date, total capacity in 1975 will be just over 105 million tons a year.

GROWTH IN INLAND CONSUMPTION OF MAIN OIL PRODUCTS

The overall growth in inland consumption of petroleum products in the past five years (26 per cent) is not very impressive in comparison with other European Member countries. This is due to the increase in the use of natural gas. The increase in consumption of each of the main oil products can be seen in the table below.

INLAND CONSUMPTION (EXCLUDING BUNKERS)

Million tons.

	1966	1967	1968	1969	1970
Motor gasoline	2.17	2.37	2.62	2.76	3.04
Kerosenes	0.98	1.03	1.14	1.19	1.23
Gas/diesel oil	4.74	5.03	5.68	5.99	6.67
Residual fuel-oil	8.56	8.11	8.13	8.33	8.05

It is expected that less fuel-oil will be used in the next few years, owing to the increasing consumption of natural gas. In the middle of the seventies, when the increase in natural gas consumption has levelled out, there will be room for growth again.

Trends in the respective shares of the different forms of energy in inland consumption (including bunkers) can be seen in the following table:

In per cent.

	1966	1967	1968	1969	1970
Coal	24	22	18	14	10
Oil	68	66	65	63.5	61
Natural gas	8	12	17	22.5	29

In the next few years the share of natural gas will continue to increase at the expense of coal and oil. Coal consumption is decreasing in absolute terms, while the decline in the case of oil is only relative.

229

NETHERLANDS

Principal Natural Gas Statistics

Billion[1] cubic metres
at 15°C and 760 mm Hg.

	1966	1967	1968	1969	1970
Indigenous production[2]	3.3	7.2	14.6	23.1	33.1
Imports	–	–	–	–	–
Total supply	3.3	7.2	14.6	23.1	33.1
Inland consumption	3.2	6.0	10.0	15.1	21.1
Exports	0.1	1.2	4.6	8.0	12.0
Total disposals	3.3	7.2	14.6	23.1	33.1

1. Billion as used in this Report is 10^9.
2. Excluding waste and flared gas, processing plant fuel, losses and return to formation.

Natural gas in the Netherlands

The importance of natural gas is illustrated inter alia by the fact that production increased tenfold during the period under review, i.e. 1966-1970. This was partly due to the sharp rise in exports, which in 1970 totalled 4 billion m³ to Germany, 4.8 billion m³ to Belgium and 3.2 billion m³ to France. Home consumption has likewise shown substantial growth, the average annual increase being some 60 per cent during the period under review.

Home consumption in 1970 included some 5 billion m³ for firing public-utility power stations, and roughly half of the remainder was by households and similar users and the other half by industry. Further sizeable increases are to be expected in home consumption and in exports over the next few years. Production will probably rise to over 80 billion m³ in the next five years. About half of this gas will be consumed at home and the remainder will be exported. By 1980, production is expected to increase to around 100 billion m³.

The steadily increasing use of natural gas for various purposes has caused a sharp increase in the share of natural gas in total primary energy consumption. In 1965 it was no more than some 5 per cent but by 1970 it had risen to 30 per cent. In the next few years it will increase to 40 or 45 per cent.

Principal Oil Statistics

Million metric tons

	1966	1967	1968	1969	1970
Crude and process oils					
Indigenous production	-	-	-	-	-
Imports	2.9	3.2	4.9	5.2	6.5
Total supply	2.9	3.2	4.9	5.2	6.5
Refinery intake	3.1	3.1	5.0	5.3	5.8
Exports	-	-	-	-	0.6
Total disposals	3.1	3.1	5.0	5.3	6.4
Petroleum products					
Refinery output	3.0	3.0	4.6	5.1	5.6
Output from other sources ...	-	-	-	-	-
Imports	4.1	4.1	3.4	3.6	4.2
Total supply	7.1	7.1	8.0	8.7	9.8
Deliveries to inland consumption	5.3	5.4	6.0	6.8	7.4
Bunkers	0.4	0.4	0.4	0.4	0.6
Exports	1.1	1.2	1.5	1.8	1.6
Total disposals	6.8	7.0	7.9	9.0	9.6
Refinery capacity					
Crude distillation capacity at end year	3.0	5.0	5.6	5.6	6.4

INDIGENOUS PRODUCTION AND IMPORTS OF CRUDE OIL

The first cargo, about 30,000 t. of crude from the oil wells in the Norwegian sector of the Continental Shelf (EKOFISK-field) arrived at Stavanger in the beginning of August 1971. It was refined at the Shell refinery for Norsk Hydro, which is a member of the Phillips Group. Since then, production has continued with a production of about 40,000 - 45,000 bbl/d. By 1st June, 1972 a total of about 7 millions of barrels has been produced from the EKOFISK-field.

It is not possible to estimate future imports of crude oil in Norway as consumption of crude oil from the Norwegian Continental Shelf is uncertain. The refinery intake, which in 1971 was estimated to be 6.0 million tons, will increase to about 8.0 million tons in 1975, within existing refineries.

DEVELOPMENT OF REFINERY CAPACITY

There are two refineries in Norway with a total capacity of about 6.4 million tons. It is estimated that the capacity will be 8.2 million tons at the end of 1972.

The industrial concern, Norsk Hydro, in co-operation with Norsk Braendselolje A/S (BP), is planning to build a refinery north of Bergen with a yearly capacity of about 4.0 million tons. It is estimated that the refinery will go on stream in 1975.

Total requirements of finished products (inland consumption and bunkers) are estimated to be 10 million tons in 1975.

GROWTH IN INLAND CONSUMPTION OF MAIN OIL PRODUCTS

The growth in inland consumption of main products from 1966 to 1970 has been:

	%
Motor gasoline	35
Jet fuels	80
Kerosenes	94
Gas/diesel oils	52
Fuel-oils	16
Total all products	41

For the period 1970 to 1975 the growth is estimated to be:

	%
Motor gasoline	34
Jet fuels	50
Kerosenes	63
Gas/diesel oils	44
Fuel-oils	7
Total all products	26

As the figures show, the estimated increase in consumption for the period 1970-1975 is more moderate than in the previous period.

Oil consumption in relation to total energy consumption

When small quantities of town gas, wood and peat (which together account for about 1 per cent of the total energy consumption) are left out Norway has the following division of energy consumption:

(In percentage)

	1966	1967	1968	1969	1970	1975
Coal and Coke	6.8	6.5	6.1	6.0	5.7	4.7
Oil products	40.4	39.0	38.3	42.3	45.7	47.2
Hydro electricity[1]	52.8	54.5	55.6	51.7	48.6	48.1
	100.0	100.0	100.0	100.0	100.0	100.0

1. 1 kWh = 0.2 kg coal equivalent. (See remarks in Annex 1, p. 5 in "Towards a new Energy Pattern in Europe" OEEC 1960).

Indigenous production of natural gas

The gas produced in association with the oil in the EKOFISK-field is for the time being flared.

A gas find of considerable size was made by the Petronord-Group in field 25, block 1 (FRIGG-field). On 25th April, 1972 A/S Petronord informed the Norwegian Ministry of Industry that all the partners in the Petronord Group considered the find to be commercially exploitable. The Norwegian State has an option to participate with 5 per cent.

Norsk Hydro participates in the group with 34.6 per cent.

The future method of transport and the landing-point of the gas produced on the Norwegian Continental Shelf has not yet been determined.

Prior to the production in the EKOFISK-field on the Norwegian Continental Shelf there was no production, import or consumption of natural gas in Norway.

233

PORTUGAL

Principal Oil Statistics[1]

Million metric tons

	1966	1967	1968	1969	1970
Crude and process oils					
Indigenous production	-	-	-	-	-
Imports	1.7	1.7	1.9	2.3	3.7
Total supply	1.7	1.7	1.9	2.3	3.7
Refinery intake	1.7	1.8	1.8	2.1	3.7
Exports	-	-	-	-	-
Total disposals	1.7	1.8	1.8	2.1	3.7
Petroleum products					
Refinery output	1.6	1.6	1.6	1.8	3.2
Output from other sources ...	-	-	-	-	-
Imports	1.4	1.6	1.9	1.9	1.5
Total supply	3.0	3.2	3.5	3.7	4.7
Deliveries to inland consumption	2.0	2.1	2.4	2.6	3.0
Bunkers[2]	0.8	0.9	0.9	1.0	1.1
Exports	0.2	0.1	0.1	0.1	0.4
Total disposals	3.0	3.1	3.4	3.7	4.5
Refinery Capacity					
Crude distillation capacity at end year	1.8	1.8	1.8	3.8	3.8

1. The figures refer to Metropolitan Portugal only.
2. Including supplies for air transport.

234

INDIGENOUS PRODUCTION AND IMPORTS OF CRUDE OIL

There is no indigenous production of crude oil in Portugal which is why it must be imported to supply the national refineries which have been working at capacity for some years, as can be seen from a comparison of the "refinery intake" and "distillation capacity" figures in the above table. (The refinery at Porto, with a 2 million tons per year capacity came on stream late in 1969, which accounts for the difference found that year between the two items.)

DEVELOPMENT OF REFINERY CAPACITY

Up to 1969 Portugal had only one refinery which was built in 1939/40 in what was then the outskirts of Lisbon but which is now in the centre of the town so that further expansion is not possible (capacity had risen from 270,000 t. in 1940 to 1,800,000 t. in 1962). At the end of 1969 the new Porto refinery, with a capacity of 2,000,000 t., came on stream. The increase in the capacity of this refinery by a further 3,000,000 t. (to 5,000,000 t./year) has already been authorised and this new capacity should come on stream in 1973. The construction of a new refinery in the southern part of Portugal has also been authorised which will have a capacity of 10 million tons/year and is planned to be built in 1975/6.

GROWTH IN INLAND CONSUMPTION OF MAIN OIL PRODUCTS

The average annual rates of increase in consumption of the main oil products over the period 1966/70 are as follows:

Liquefied petroleum gas (LPG) .... + 14.9 %
Motor gasoline .... + 11.2 %
Kerosene .... — 7.1 %
Gas-oil: excluding "bunkers" .... + 9.8 %
      including "bunkers" .... + 8.2 %
Fuel-oil (not including that used for electricity gene-
    ration):
      excluding "bunkers" .... + 6.6 %
      including "bunkers" .... + 6.5 %
Fuel-oil for electricity generation .... + 53.3 %
Jet fuel (JP4 and JP1) .... + 18.2 %

Disregarding jet fuel, where the increase in consumption is accounted for exclusively by the growth in international air traffic, and fuel-oil used for electricity generation, liquefied petroleum gas shows the highest rate of increase. Liquefied petroleum gas is a substitution fuel replacing solid fuels for household use or even kerosene which is showing a very steep downward trend (— 7.1 per cent).

The increase in the consumption of gasoline (+ 11.2 per cent) is in line with the growth in the vehicle population and is also partly due to the development of tourism.

The increase in the use of gas-oil is largely the result of agricultural mechanisation and the diesel engines which are practically exclusively used.

The high figure (53.3 per cent) for fuel-oil used for electricity generation is explained by the fact that the first oil-fired thermal power station came into service in 1969. Since its purpose is to act as a reserve for hydro-electric power its annual consumption figures are very variable.

OIL CONSUMPTION IN RELATION TO TOTAL ENERGY CONSUMPTION

Oil consumption as a percentage of the consumption of all forms of energy has been as follows:

1966	61 per cent
1967	62 per cent
1968	65 per cent
1969	63 per cent
1970	71 per cent

Portugal has very limited coal resources. The rainfall pattern is extremely variable and in view of the fact that hydraulic resources are practically all harnessed (in satisfactory economic conditions) hydro-electric generation would have great difficulty in helping to meet future increases in energy requirements.

No contribution can come from nuclear energy before the end of the present decade (because of the time required to build power stations), and therefore oil products (and possibly natural gas) will have to meet practically the entire increase in energy requirements, at least up to about 1980.

NATURAL GAS

Natural gas is neither produced nor consumed in Portugal.

SPAIN

Principal Oil Statistics

Million metric tons

	1966	1967	1968	1969	1970
Crude and process oils					
Indigenous production	–	–	–	–	–
Imports	15.9	21.2	27.5	29.1	32.2
Total supply	15.8	21.1	27.5	29.1	32.2
Refinery intake	15.8	21.3	27.2	29.1	32.2
Exports	–	–	–	–	–
Total disposals	15.8	21.3	27.2	29.1	32.2
Petroleum products					
Refinery output	15.5	19.8	26.6	28.9	32.3
Output from other sources ...	–	–	–	–	–
Imports	0.9	0.8	0.8	1.1	1.3
Total supply	16.4	20.6	27.4	30.0	33.6
Deliveries to inland consumption	12.5	15.8	19.5	22.9	25.2
Bunkers	0.7	1.6	1.7	2.0	2.0
Exports	0.5	2.6	5.8	5.2	5.4
Total disposals	13.7	20.0	27.0	30.1	32.6
Refinery capacity					
Crude distillation capacity at end year	19.8	28.4	28.4	30.8	34.8

SPAIN

Principal Natural Gas Statistics

Billion[1] cubic metres
at 15°C and 760 mm Hg.

	1966	1967	1968	1969	1970
Indigenous production[2]	-	-	-	-	-
Imports	-	-	-	-	15.1
Total supply	-	-	-	-	15.1
Inland consumption	-	-	-	-	9.6
Exports	-	-	-	-	-
Total disposals	-	-	-	-	9.6

1. Billion as used in this Report is 10^9.
2. Excluding waste and flared gas, processing plant fuel, losses and return to formation.

Principal Oil Statistics

Million metric tons

	1966	1967	1968	1969	1970
Crude and process oils					
Indigenous production	-	-	-	-	-
Imports	4.1	6.5	9.4	11.0	11.9
Total supply	4.1	6.5	9.4	11.0	11.9
Refinery intake	4.0	6.3	9.5	11.0	11.9
Exports	-	-	-	-	-
Total disposals	4.1	6.3	9.5	11.0	11.9
Petroleum products					
Refinery output	3.7	5.9	8.8	10.3	11.4
Output from other sources ...	0.1	0.1	-	-	-
Imports	18.5	16.3	16.9	18.4	20.6
Total supply	22.3	22.3	25.7	28.7	32.0
Deliveries to inland consumption	20.0	19.5	22.4	25.2	28.2
Bunkers	0.9	1.1	1.2	1.2	1.2
Exports	0.4	0.6	1.2	1.7	1.8
Total disposals	21.3	21.2	24.8	28.1	31.2
Refinery capacity					
Crude distillation capacity at end year	6.0	11.3	12.6	12.6	12.9

Indigenous production and imports of crude oil

There has been no production of crude oil in Sweden up to now.

A Swedish company has however been searching for oil and gas for some time. There have been test drillings in the South of Sweden and in the future test drillings might be undertaken on the Continental Shelf as well.

As yet no positive results have been achieved. A cost limit of SKr 100 million has been set for the first five years of exploration. If there are positive indications the period of exploration might be extended to eight years and the cost limit to SKr 150 million.

Sweden's yearly imports of crude oil, feedstocks and components amounted to about 4 million tons in 1966 and have been trebled from 1966 to 1970, when more than 12 million tons were imported. In 1966, 40 per cent of these imports came from the Middle East, 20 per cent from Africa and 35 per cent from Venezuela, the remaining 5 per cent being imports of feedstocks from the OECD area. The corresponding figures for 1970 are 48 per cent from the Middle East, 30 per cent from Africa, 16 per cent from Venezuela and 6 per cent from the Soviet Union.

Development of refinery capacity

Refinery capacity in Sweden which was about 6 million tons in 1966 has been doubled during the period to 12 million tons. In the next few years this capacity will rise by 7 million tons through the construction of a new refinery. There are also plans for considerable enlargements of the existing refineries. If these plans are realised the refinery capacity in Sweden can be expected to reach 35-40 million tons in 1980.

Growth in inland consumption of main oil products

Inland consumption of oil products has risen from 20 million tons in 1966 to over 28 million tons in 1970 i.e. by 40 per cent. A study of the development from year to year shows that total consumption was lower in 1967 than in 1966, which was mainly due to better conditions for production of hydro power and a less cold winter.

Compared to a normal winter, 1966 was 11 per cent colder whereas 1967 was 4 per cent warmer. The increase of oil consumption during 1968, 1969 and 1970 is mainly due to the considerable rise in consumption of oil for electric power production. The table below shows development of consumption with index 100 for 1966.

	1966 MILLION TONS	1967	1968	1969	1970
		INDEX			
Motor gasoline	2.3	104	109	115	120
Gas/diesel oils	7.1	96	108	116	127
(of which heating)	(4.7)	(97)	(110)	(118)	(127)
Fuel oil	9.1	93	117	132	156
(of which heating)	(3.5)	(100)	(107)	(110)	(121)
(of which power plants) ..	(0.7)	(29)	(140)	(318)	(553)

In the foreseeable future a continued increase of oil consumption is expected although with certain variations between the years due to weather conditions and the amount of water available for hydro power production. According to present forecasts total consumption of oil products is expected to have reached 33 million tons in 1975 and about 40 million tons in the middle eighties.

OIL CONSUMPTION IN RELATION TO TOTAL ENERGY CONSUMPTION

Of total energy consumption in Sweden in 1966, hydro power accounted for 14 per cent (1 kWk = 860 kcal), wood and pulp lyes for 9 per cent, coal and coke for 7 per cent and oil products for 70 per cent. In 1970 the relations were 10 per cent hydro power, 8 per cent wood and pulp lyes, 4 per cent coal and coke and 78 per cent oil products. If hydro power conditions had been normal in 1970 the relation would have been about 75 per cent oil products and 13 per cent hydro power. Assuming that the Swedish nuclear power programme is carried out according to current plans, the share of oil products in the total energy supply is expected to decline from the late seventies. Even thereafter a continued increase in the volume of oil consumption is expected.

NATURAL GAS

Natural gas is neither produced nor consumed in Sweden.

SWITZERLAND

Principal Oil Statistics

Million metric tons

	1966	1967	1968	1969	1970
Crude and process oils					
Indigenous production	-	-	-	-	-
Imports	2.4	4.0	4.6	5.1	5.5
Total supply	2.4	4.0	4.6	5.1	5.5
Refinery intake	2.3	4.0	4.6	5.1	5.5
Exports	-	-	-	-	-
Total disposals	2.3	4.0	4.6	5.1	5.5
Petroleum products					
Refinery output	2.2	3.7	4.3	4.8	5.2
Output from other sources ..	-	-	-	-	-
Imports	6.8	5.6	6.3	6.8	7.8
Total supply	9.0	9.3	10.6	11.6	13.0
Deliveries to inland consumption	8.3	9.1	10.0	11.0	12.3
Bunkers	-	-	-	-	-
Exports	0.1	0.2	0.2	0.3	0.3
Total disposals	8.4	9.3	10.2	11.3	12.6
Refinery capacity					
Crude distillation capacity at end year	4.6	4.7	5.1	5.1	5.5

242

INDIGENOUS PRODUCTION AND IMPORTS OF CRUDE OIL

There is no indigenous production of crude oil in Switzerland. In spite of much exploration work and a number of test drillings no economically exploitable field has been discovered. Up to now, all exploration has been carried out by private entreprise, the Government taking no initiative whatsoever in this field. Efforts are being made at the present time to continue the search for oil, possibly with financial assistance from the authorities.

All imports of crude oil into Switzerland are by pipeline from the Mediterranean ports. When the second refinery came on stream in 1966 imports more than doubled. There is a clear cut preference for Algerian and Libyan crude, and especially for the latter in recent years. This is because the refiners in Switzerland are looking for maximum distillate yield, by far the highest demand being for heating oil.

DEVELOPMENT OF REFINERY CAPACITY

Up to mid-1956, Raffinerie du Rhône S. A., which came on stream in 1963 with an annual capacity of 2.1 million tons, was the only refinery in the country. In June 1966 the Shell refinery came on stream, bringing the country's total capacity up to 4.6 million tons a year. As a result of technical improvements and rationalisation output rose to 5.1 million tons in 1968 and 5.5 million tons in 1971. Further expansion is planned to bring the total refinery capacity in Switzerland up to about 6.5 million tons a year.

Since the consumption of oil products and the demand for all energy sources are constantly rising, it is necessary to give further thought to the questions of a third refinery in Switzerland. This is still no more than a project at the present time but, in view of the fact that only a small fraction of current inland consumption is covered by the existing two refineries, sooner or later it will be essential to build a third.

GROWTH IN INLAND CONSUMPTION OF MAIN OIL PRODUCTS

Total sales of oil products increased by 47 per cent between 1966 and 1970.

Of all oil products, the greatest increase has been in sales of jet fuels (nearly 80 per cent) due to the growth of international air traffic.

Deliveries of distillates have increased by 57 per cent. This is due to intensive housing construction and the conversion of many domestic heating installations from coal to oil. The same trend is also to be seen in the case of industrial fuel-burning equipment and, coupled with the high level of industrial activity, it has contributed to the big increase (nearly 50 per cent) in the consumption of fuel-oil.

Motor fuel sales (motor gasolines and diesel oil) show an average annual growth rate of 6.5 per cent, making 26 per cent over the full period.

The possible repercussions of the present monetary crisis and the signs of an approaching economic recession make it difficult to forecast future growth in the consumption of oil products. It is, however, expected that demand will continue to increase but at a slower rate than in recent years because other forms of energy are assuming greater importance on the Swiss energy market.

OIL CONSUMPTION IN RELATION TO TOTAL ENERGY CONSUMPTION

Over the period 1966 to 1970, the share of oil products in total energy consumption has increased appreciably at the expense of other primary energy sources. From 70.2 per cent in 1966, it rose to 78.4 per cent in 1970, an average annual increase of 2.1 per cent.

The share of motor fuels in total energy consumption remained initially unchanged but that of other liquid fuels has increased steadily from 48.4 per cent in 1966 to 56.4 per cent in 1970.

SWITZERLAND

Principal Natural Gas Statistics

Billion[1] cubic metres
at 15°C and 760 mm Hg.[3]

	1966	1967	1968	1969	1970
Indigenous production[2]	–	–	–	–	–
Imports	–	–	–	0.00025	0.014
Total supply	–	–	–	0.00025	0.014
Inland consumption	–	–	–	0.00025	0.014
Exports	–	–	–	–	–
Total disposals	–	–	–	0.00025	0.014

1. Billion as used in this Report is 10^9.
2. Excluding waste and flared gas, processing plant fuel, losses and return to formation.
3. Calorific value: 8,400 Kcal/cu.m.

INDIGENOUS PRODUCTION AND IMPORTS OF NATURAL GAS

So far the search for natural gas in Switzerland has been unsuccessful. A small field discovered near Pfaffnau proved not to be economically viable. Hope has not been abandoned altogether, however, and plans for exploration in the Alpine region are currently receiving serious consideration.

Natural gas become available in Switzerland only very recently, namely in 1969, when small quantities appeared on the Swiss market from Pfullendorf (Federal Republic of Germany), north of Lake Constance.

Major projects are now in hand. As from 1974 Switzerland will be in a position to take 500 million m³ of natural gas per annum from the Netherlands-Italy pipeline which will pass through the country.

Concession for the laying of the Netherlands-Italy transit pipeline has been granted on 2nd February 1972. To supplement these deliveries of Dutch gas and the relatively small quantities coming in from Pfullendorf, the gas industry is currently negotiating arrangement for further supplies from other sources. Natural gas is already imported from the Netherlands by the "Communauté du Gaz du Mittelland, S. A.", via France, through the Schönenbuck-Arlesheim pipeline.

245

Inland consumption of natural gas

Natural gas is now becoming available on the Swiss gas market on a very small scale. In 1970 it was only of marginal significance in total gas consumption which amounted to nearly 500 million m³ of a 4,200 kcal/m³ calorific value.

About 80 per cent of the gas consumed in Switzerland is made in the country, one-third being from coal and two-thirds from oil products. Out of the 20 per cent that is imported, natural gas so far accounts for only 6 per cent.

Consumption of natural gas in relation to total energy consumption

The share of natural gas is still very small. All told, imported gas (natural gas and town gas) accounted for only 0.1 per cent of the country's total energy consumption in 1968 and 1969, and 0.3 per cent in 1970.

Natural gas will not begin to play a more important part in Switzerland's energy balance until 1974/1975.

Principal Oil Statistics

Million metric tons

	1966	1967	1968	1969	1970
Crude and process oils					
Indigenous production	2.1	2.7	3.1	3.6	3.5
Imports	3.1	3.0	3.4	2.9	3.9
Total supply	5.2	5.7	6.5	6.5	7.4
Refinery intake	5.0	5.5	6.4	6.5	7.2
Exports	-	-	-	-	-
Total disposals	5.0	5.5	6.4	6.5	7.2
Petroleum products					
Refinery output	4.8	5.3	6.3	6.4	7.0
Output from other sources ...	-	-	-	-	-
Imports	0.2	0.2	0.4	0.7	0.6
Total supply	5.0	5.5	6.7	7.1	7.6
Deliveries to inland consumption	4.0	4.3	5.8	6.6	6.6
Bunkers	-	-	-	-	-
Exports	0.5	0.2	0.3	0.1	0.1
Total disposals	4.5	4.5	6.1	6.7	6.7
Refinery capacity					
Crude distillation capacity at end year	5.2	5.6	6.5	6.5	7.5

INDIGENOUS PRODUCTION AND IMPORTS OF CRUDE OIL

As it is indicated in the Principal Oil Statistics, indigenous production of crude oil has been increasing with the exception of the year 1970 for which a small drop is observed. Efforts are being made to increase the indigenous production and reserves of crude oil. On the other hand, imports of crude oil are also increasing and it is expected they will increase further in the near future with the addition of new refineries.

DEVELOPMENT OF REFINERY CAPACITY

Refinery capacity is also expected to increase.

GROWTH IN INLAND CONSUMPTION OF MAIN OIL PRODUCTS

Inland consumption of main oil products was about constant at the beginning of the period 1966-1970 but toward the end of this period an increase of one million tons is seen.

OIL CONSUMPTION IN RELATION TO TOTAL ENERGY CONSUMPTION

Oil consumption made up 33.9 per cent of the total energy consumption of Turkey in 1970.

NATURAL GAS

Natural gas is neither produced nor consumed in Turkey.

UNITED KINGDOM

Principal Oil Statistics

Million metric tons

	1966	1967	1968	1969	1970
Crude and process oils					
Indigenous production	0.1	0.1	0.1	0.1	0.2
Imports	71.5	73.7	83.2	93.2	102.1
Total supply	71.6	73.8	83.3	93.3	102.3
Refinery intake	71.7	73.5	83.1	91.7	101.9
Exports	0.1	0.7	0.1	0.3	1.2
Total disposals	71.8	74.2	83.2	92.0	103.1
Petroleum products					
Refinery output	66.3	68.0	77.1	85.1	94.7
Output from other sources ...	0.4	0.3	0.3	0.2	0.2
Imports	21.7	23.9	22.6	20.6	20.1
Total supply	88.4	92.2	100.0	105.9	115.0
Deliveries to inland consumption	69.1	74.6	79.0	85.1	90.8
Bunkers	5.0	5.1	5.3	5.6	5.5
Exports	12.6	11.9	14.5	14.3	17.4
Total disposals	86.7	91.6	98.8	105.0	113.7
Refinery capacity					
Crude distillation capacity at end year	83.8	86.1	97.6	109.4	114.7

249

Indigenous production and imports of crude oil

The United Kingdom's oil requirements were met almost entirely by imports in the period 1966 to 1970; crude oil imported rose in this period from 71.5 million tons in 1966 to 102.2 million tons in 1970. On average oil producing countries in the Middle East supplied 59 per cent of imports.

Only minor amounts of indigenous crude oil were produced from 1966 to 1970. Commercially significant oil fields have however recently been discovered in the United Kingdom continental shelf beneath the North Sea. The reserves already found are believed to be sufficient to sustain a production of 25 million tons a year within the next few years and perhaps treble this amount by the end of the decade. Production of natural gas liquids from offshore gas fields began in 1967 and in 1970 accounted for about half the indigenous production of liquid petroleum.

Development of refinery capacity

During 1966-70 refinery capacity was increased broadly in step with the growth in consumption. At the end of 1970 refinery capacity was 37 per cent greater than at the end of 1966. Notable additions to capacity include a 5.4 million ton refinery at Teesport, a 7.1 million ton refinery at South Killingholme, a 4.1 million ton refinery at Milford Haven, a 4.1 million ton refinery at Killingholme and a 4.6 million ton expansion of capacity at Grangemouth. Known projects will add 39 million tons to refining capacity by the end of 1975.

Growth in inland consumption of oil products

Inland consumption of oil products increased from 69 million tons in 1966 to 90.8 million tons in 1970. Consumption of fuel-oil rose by 31 per cent between 1966 and 1970; the corresponding increases for gas/diesel oil and motor gasoline were 34 per cent and 24 per cent respectively. In 1970 the proportions of inland consumption of products accounted for by fuel-oil, gas/diesel oil and motor gasoline were 42 per cent, 19 per cent and 16 per cent respectively, virtually the same as in 1966.

Oil consumption in relation to total energy consumption

Total inland energy consumption in the United Kingdom rose by 10.6 per cent between 1966 and 1970. Consumption of petroleum for energy increased by 30.4 per cent and in 1970 petroleum accounted for 44.4 per cent of total energy consumption compared to 37.5 per cent in 1966.

UNITED KINGDOM

Principal Natural Gas Statistics

Billion[1] cubic metres
at 15°C and 760 mm Hg.

	1966	1967	1968	1969	1970
Indigenous production[2]	0.17	0.62	2.19	5.23	11.21
Imports	0.69	0.81	1.08	1.08	0.91
Total supply	0.86	1.43	3.27	6.31	12.12
Inland consumption	0.86	1.43	3.27	6.29	12.08
Losses	-	-	-	0.02	0.04
Exports	-	-	-	-	-
Total disposals	0.86	1.43	3.27	6.31	12.12

1. Billion as used in this Report is 10^9.
2. Excluding waste and flared gas, processing plant fuel, losses and return to formation.

INDIGENOUS PRODUCTION AND IMPORTS OF NATURAL GAS

Liquefied natural gas has been imported from Algeria since 1964 but from 1968 indigenous production has been of greater importance. Production began on a significant scale in 1967 when gas from the West Sole field beneath the North Sea was first pumped ashore. One mainland and three offshore fields are now supplying gas for inland consumption. Production from the Indefatigable offshore field is due to commence towards the end of 1971 and from the Viking offshore field in 1972. By 1975 annual production is expected to exceed 40 billion cubic metres.

INLAND CONSUMPTION OF NATURAL GAS

Inland consumption of natural gas virtually doubled in each of the years 1967 to 1970 and in the latter year exceeded 12 billion cubic metres. In 1970 two-thirds of the natural gas available to the gas industry was used in the manufacture of town gas and the balance was supplied direct to consumers. By 1975 natural gas will meet almost all the raw material requirements of the gas industry.

CONSUMPTION OF NATURAL GAS IN RELATION TO TOTAL ENERGY CONSUMPTION

In 1966 natural gas accounted for only 0.4 per cent of the total energy requirements of the United Kingdom. By 1970 this proportion had increased to 4.9 per cent.

UNITED STATES

Principal Oil Statistics

Million metric tons

	1966	1967	1968	1969	1970
Crude and process oils					
Indigenous production[1]	447.9	477.1	501.3	509.7	532.6
Imports	65.0	61.3	68.7	75.6	71.4
Total supply	512.9	538.4	570.0	585.3	604.0
Refinery intake[1]	490.7	509.5	539.9	557.2	570.9
Exports	0.2	3.6	0.2	0.2	0.7
Total disposals	490.9	513.1	540.1	557.4	571.6
Petroleum products					
Refinery output	488.9	507.9	535.7	550.5	565.5
Output from other sources[2] ..	23.6	23.9	31.3	33.3	35.6
Imports	72.1	69.7	77.6	86.9	104.2
Total supply	584.6	601.5	644.6	670.7	705.3
Deliveries to inland consumption	563.8	574.9	613.7	645.7	681.6
Bunkers	8.7	9.2	9.6	15.4	8.4
Exports	10.0	12.1	11.9	12.0	13.5
Total disposals (including statistical differences)	584.5	601.5	644.6	670.7	705.3
Refinery capacity					
Crude distillation capacity at end year[3]	10,760	11,533	11,740	12,074	13,020

1. Including natural gas liquids.
2. Products produced at natural gas processing plants.
3. In thousands of barrels per calendar day.

CONSUMPTION

Domestic consumption of petroleum products increased from 3,536 million barrels in 1960 to 5,245 million in 1970, for an average annual growth rate of 4.0 per cent in the ten-year period. Petroleum products, including natural gas liquids, supplied 44.8 per cent of the United States energy requirements in 1960 but this contribution declined to 43.4 per cent in 1970 reflecting the increased use of natural gas, principally in the residential heating market.

The following table shows consumption of the major petroleum products and the share of the total that each contributed for the years 1960 through 1970. Liquefied petroleum gases and jet fuels represent a large portion of the "Other" category and the consumption of these fuels has expanded rapidly in the past ten years (jet fuel at an annual rate of almost 10 per cent, and liquefied gases at about 7 per cent). The table also illustrates the significant relative change in distillate and kerosene resulting from the competition of natural gas. Since 1968, residual fuel-oil has been making a strong comeback in the electric power generation and industrial markets as a result of air quality regulations adopted in several State and metropolitan areas. Supplies of coal with a low sulphur content are presently limited; the natural gas industry has had to limit its deliveries to utility and industrial consumers to meet the growing demand in other market sectors; and delays have been encountered in bringing nuclear power plants into operation.

Motor gasoline demand for the next five years is expected to increase at the rate of 3.3 per cent per year compared with the previous five-year average annual growth rate of 4.7 per cent, while the demand for aviation fuels is expected to continue increasing at about the same annual growth rate (7.4 per cent) as for the past five-year period.

Because of air quality regulations adopted in several States and metropolitan areas, and the limited supplies of natural gas available to their markets, electric utility and industrial users will be more dependent on distillate fuel and low-sulfur residual fuel-oil. This trend should continue until effective and inexpensive methods are developed to eliminate objectional stack gas emissions from coal burning facilities.

Petroleum including natural gas liquids represented 43.4 per cent of total energy demand in the United States in 1970. This share is expected to decline to 40.8 per cent by 1975 when additional nuclear power facilities are expected to be in operation.

EXPORTS

Exports reached a low of 61 million barrels in 1962 following several years of decline, but since that time have increased to the 1970 level of 94 million barrels. Petroleum coke, lubricating oils and greases, and residual fuel-oil account for about 75 per cent of the exported products.

IMPORTS

Imports in 1960 represented 17 per cent of total supply but by 1970 this had increased to 23 per cent. Crude oil imports increased at an average annual rate of 2.7 per cent while the increase for refined products

253

CONSUMPTION - MAJOR PRODUCT CATEGORIES

Million barrels

	TOTAL PRODUCTS	MOTOR GASOLINE[1]	DISTILLATE INCLUDING KEROSENE	RESIDUAL FUEL OIL	OTHER
1962	3,736	1,533	896	545	762
Percentage	100.0	41.0	24.0	14.6	20.4
1963	3,851	1,582	919	539	811
Percentage	100.0	41.1	23.9	14.0	21.0
1964	3,959	1,611	929	555	864
Percentage	100.0	40.7	23.5	14.0	21.8
1965	4,125	1,676	873[2]	587	989
Percentage	100.0	40.6	21.2	14.2	24.0
1966	4,325	1,755	874	626	1,070
Percentage	100.0	40.6	20.2	14.5	24.7
1967	4,481	1,810	919	652	1,100
Percentage	100.0	40.4	20.5	14.6	24.5
1968	4,789	1,925	977	668	1,219
Percentage	100.0	40.2	20.4	13.9	25.5
1969	5,042	2,017	1,001	722	1,302
Percentage	100.0	40.0	19.9	14.3	25.8
1970 p/	5,245	2,111	1,023	804	1,307
Percentage	100.0	40.3	19.5	15.3	24.9

1. Excluding aviation gasoline.
2. Beginning with 1965, excludes kerosene-type jet fuel.
p/ Preliminary.

254

was 4.9 per cent. During this ten-year period there have been several changes in the oil import regulations which have relaxed import restrictions for refined products, primarily in the East Coast market. Deep water terminal operators are now licensed to import all the residual fuel-oil and heavy distillate fuel-oil necessary to meet their customer requirements and special provisions have been granted to refiners in Puerto Rico and the Virgin Islands to ship a part of their refined product output to the East Coast. Among other changes and amendments natural gas liquids produced in other Western Hemisphere nations are now eliminated from the import control restrictions. The Canadian share of the United States import market has expanded rapidly since 1960, from 6.6 per cent to 22.4 per cent, while the share for other Western Hemisphere countries has concomitantly declined from 73.8 per cent to 61.2 per cent in 1970. The United States imported 245.3 million barrels of crude oil from Canada in 1970 compared with just 41.3 million in 1960.

REFINERY ACTIVITY

As of January 1971, petroleum refineries were operating at 86 per cent of crude throughput capacity and had a total installed crude oil capacity of 13,020 thousand barrels per day. Completion of additional capacity now under construction is expected to increase this capacity at the end of 1971 by almost 5 per cent.

Concern for the environment, the uncertainty of future oil import policies and consideration of a variety of economic factors are clearly reflected in the relative scarcity of firm new oil refinery construction projects in the United States. Planned expansion of existing facilities and the building of new refineries when placed against future petroleum product requirements, fall short of giving the United States the refining capacity it will need by 1975.

PRODUCTION

Over the past ten years crude oil production increased at an average rate of 3.2 per cent per year and reached 3,517.5 million barrels in 1970. Twice during this period the United States has had to rely on spare production capacity to offset the loss of imports from overseas. The first was during the Middle East crisis of 1967 which closed the Suez Canal and the second occurred in 1970 when the pipeline from the Middle East to the Meditarranean was shut down and Libya curtailed production. Natural gas liquids production in 1970 was 605.9 million barrels. Since 1960 production of this commodity has increased at the rate of 6.6 per cent per year.

The United States production of crude oil and natural gas liquids has increased at the rate of 4.5 per cent per year for the past four years; however, the demand for products has increased at a rate of 4.8 per cent, thus increasing reliance has had to be placed on foreign oil to fill the supply-demand deficit. A growth rate of about 2.3 per cent is expected for indigenous production other than Alaska for the next five years while product demand growth is expected to increase at an average annual rate of four per cent.

SOURCES OF US IMPORTS OF PETROLEUM

Million barrels

	1962	1963	1964	1965	1966	1967	1968	1969	1970p)
Western Hemisphere									
Crude	267.7	278.9	289.0	283.4	289.8	301.0	316.0	337.6	351.4
Products	338.1	352.9	375.4	434.6	467.8	481.7	513.0	566.7	691.5
Total	605.8	631.8	664.4	718.0	757.6	782.7	829.0	904.3	1,042.9
Percent of total:									
Crude	65.1	67.6	65.9	62.7	64.8	73.1	66.9	65.7	72.7
Products	96.9	97.5	96.7	96.9	95.1	93.7	90.5	88.4	90.5
Total	79.7	81.5	80.4	79.7	80.7	84.5	79.8	78.3	83.6
Eastern Hemisphere:									
Crude	143.4	133.8	149.6	168.6	157.3	110.6	156.3	176.5	131.9
Products	10.6	9.2	12.7	14.1	24.2	32.6	54.0	74.7	72.6
Total	154.0	143.0	162.3	182.7	181.5	143.2	210.3	251.2	204.5
Percent of total:									
Crude	34.9	32.4	34.1	37.3	35.2	26.9	33.1	34.3	27.3
Products	3.1	2.5	3.3	3.1	4.9	6.3	9.5	11.6	9.5
Total	20.3	18.5	19.6	20.3	19.3	15.5	20.2	21.7	16.4
World:									
Crude	411.1	412.7	438.6	452.0	447.1	411.6	472.3	514.1	483.3
Products	348.7	362.1	388.1	448.7	492.0	514.3	567.0	641.4	764.1
Total	759.8	774.8	826.7	900.7	939.1	925.9	1,039.3	1,155.5	1,247.4

p) Preliminary.

TRANSPORTATION

Pipelines are the principal means of transporting crude oil and refined products in the United States. The second major movement are the tankers and barges used in the coastal movements from the Gulf Coast to the East Coast. River barges and rail shipments also are a part of the transportation systems utilized to move the crude oil and products. Final deliveries of the finished products to the consumers are usually by tank truck.

UNITED STATES
Principal Natural Gas Statistics

Billion[1] cubic metres
at 15°C and 760 mm Hg.

	1966	1967	1968	1969	1970
Indigenous production[2]	487.2	514.5	547.1	586.1	620.7
Imports	13.6	16.0	18.5	20.6	23.2
Total supply	500.8	530.5	565.6	606.7	643.9
Inland consumption	486.7	514.6	551.0	592.4	624.2
Exports	0.7	2.3	2.7	1.4	2.0
Total disposals	487.4	516.9	553.7	593.8	626.2

1. Billion as used in this Report is 10^9.
2. Excluding waste and flared gas, processing plant fuel, losses and return to formation.

NATURAL GAS

Since 1960, consumption of natural gas has increased at an annual rate of almost 6 per cent and this fuel now supplies about a third of the energy requirements of the United States. During the past several years there has been great concern over future supply as domestic production has exceeded the finding of new reserves. Although there are 26,000 billion cubic feet of proved reserves of natural gas in the northern area of Alaska it will be several years before pipeline facilities can be built to move this gas to the consuming markets.

The indigenous production of natural gas in the United States is expected to increase almost 25 per cent by 1975 in response to rapidly rising requirements and higher wellhead prices. Pipeline imports of natural gas are expected to double or possibly even triple by the same date in order to fill the needs not met by indigenous production. Research efforts to develop synthetic gas from coal are not likely to produce significant supplies by 1975 and imports of liquefield natural gas are not expected to become important until the latter part of the decade. Some companies are considering the manufacture of synthetic gas from petroleum; this could be a significant source of gas for selected local markets by the mid-1970's.

Inland consumption of natural gas in the United States is expected to increase by almost 27 per cent by 1975 as a result of the increasing importance of low-sulfur fuels in achieving environmental quality standards. This growth could be expected to be even greater if the anticipated tight supply situation were alleviated.

The portion of total energy consumption presently accounted for by natural gas is expected to drop only slightly from the present 33 per cent by 1975.

STATISTICALS TABLES

NOTES AND DEFINITIONS

NOTE

The main purpose of this Annex is to present national statistics of interest which would be out of place in the text of the report, and to give some additional statistics which may be useful for reference purposes.

SOURCES OF THE STATISTICS

a) OECD Europe

Most of the statistics used in the text and appearing in the tables of this Annex have been taken from OECD sources.

b) United States and Canada

Most of the statistics have been derived from United States and Canadian official publications.

c) World statistics

These statistics have been compiled from various official and industry sources.

d) Country chapters in Annex I

These statistics were furnished by the respective Member countries.

DEFINITIONS

1 metric ton	1,000 kilograms = 0.985 ton = 2,205 pounds*
1 toe	1 ton of oil equivalent**
1 Mtoe	1 million tons of oil equivalent
1 cubic metre	35.3 cubic feet
1 billion	1,000 million = 10^9
1 trillion	1,000 billion = 10^{12}
fob	free on board
cif	cost including insurance and freight
dwt	dead weight ton
GNP	Gross National Product
n.a.	not available

* Metric tons have been used throughout this report.
** The following coefficients have been used.

Quantities	Energy content in million tons of oil equivalent (1 Mtoe = 10^{13} kcal = 4 × 10^{13} BTU)
1 million tons crude oil or petroleum products	1.0
1 million tons coal	0.7
1 million tons coke	0.67
1 million tons patent fuel	0.7
1 million tons lignite	0.2
1 million tons lignite briquettes	0.48
1,000 million m³ natural gas	0.9
1,000 million m³ manufactured gas	0.42
1 TWh electricity (units sent out and final consumption)	0.086

1 BTU .... the quantity of heat required to raise the temperature of one pound of water 1 degree F

1 kilometre ...... 1,000 metres = 0.62137 miles

1 metre .... 39.37 inches = 3.28 feet

«North America» = Canada and the United States.

«OECD Europe» or «Western Europe» = Austria, Belgium, Denmark, Finland, France, Federal Republic of Germany, Greece, Iceland, Ireland, Italy, Luxembourg, the Netherlands, Norway, Portugal, Sweden, Spain, Switzerland, Turkey and the United Kingdom.

CONVERSION FACTORS

Approximate Conversion Factors

(Referring to an average petroleum product of about 25/26 API gravity, or about 0.900 specific gravity).

To convert:
Multiply by:

metric tons to US gallons	300
metric tons to barrels (bbls)	7
metric tons per year to bbls per day	0.02

Precise Conversion Factors

To convert:
Multiply by:

metric tons to long tons 0.98421

metric tons to barrels:

a) for aviation gasoline (API gravity: 72; specific gravity: 0.695)	9.074
b) for motor gasoline (API gravity: 63.5; specific gravity: 0.725)	8.691
c) for kerosene (API gravity: 43.2; specific gravity: 0.810)	7.776
d) for gas/diesel oil (API gravity: 33.5; specific gravity: 0.857)	7.353
e) for lubricating oil (API gravity: 25.7; specific gravity: 0.900)	7.004
f) for heavy fuel oil (API gravity: 20.5; specific gravity: 0.930)	6.780

To convert US barrels of 42 gallons in metric tons, divide by:

Crude oil

Algeria-Sahara	7.713
Angola	7.223
Bahrain	7.355
British Borneo	7.709
Canada	7.428
Colombia	7.054
Egypt	6.849
Gabon	7.245
Indonesia	7.348
Iran	7.370
Iraq	7.452
Kuwait	7.263
Libya	7.593
Mexico	7.104
Neutral zone	6.849
Nigeria	7.361
Qatar	7.709
Romania	7.453
Saudi Arabia	7.370
Trinidad	6.989
United States	7.418
USSR	7.096
Venezuela	7.005
Other countries not specified (average factor)	7.3

Natural gas liquids 10.919

Refined products

Liquefied petroleum gas	11.80
Aviation gasoline	8.90

Gasoline type jet fuel 8.27
Kerosene type jet fuel 7.93
Motor gasoline and naphta 8.53
Kerosenes 7.73
Unfinished oils — Feedstocks — Light distillates for gas working 7.46
Gas diesel oil 7.46
Residual fuel oil 6.66
Lubricants 7.06
Bitumen — Asphalt — Road oil 6.10
Paraffin waxes 7.87
Petroleum coke 5.50
Benzol 7.14
Petrolatum 7.87
Miscellaneous 8.00

* *

Refinery gas

To convert millions of cubic feet in metric tons, divide by 39.729

Natural gas

To convert cubic feet in cubic metres, multiply by 0.028317

«Eastern Europe» = the Soviet Union, Albania, Bulgaria, Czechoslavakia, East Germany, Hungary, Poland, Romania and Yugoslavia. The latter seven countries make up «Eastern Europe» when the term «USSR and Eastern Europe» is employed.

«Communist countries» = Eastern Europe plus the Chinese People's Republic, North Korea, North Vietnam and Cuba.

CONTENTS

Tables

1. Primary energy consumption by form of energy, OECD area, 1960 and 1970 265
2. Per capita inland consumption of petroleum products, OECD area, 1966-1970 266
3. Refinery output and consumption (including bunkers) of petroleum products, OECD area, 1966 and 1970 267
4. Refinery output and consumption (including bunkers) of main petroleum products by countries, OECD area 1970 268
5. Estimated principal end-uses of oil products, OECD European area, North America and Japan, 1966-1970 270
6. Crude oil production, OECD area, 1966-1970 271
7. Sources of imports of crude, feedstocks and components and petroleum products, OECD area, 1966-1970 272
8. Sources of imports of crude oil, feedstocks and components by countries, OECD area, 1970 276
9. Sources of imports of petroleum products by countries, OECD area, 1970 278
10. Exports of petroleum products with destination and annual tonnage of bunkers, OECD area, 1966-1970 279
11. OECD area exports of petroleum products by destination, 1970 281
12. Intra-OECD area trade in petroleum products, 1970 282
13. Net imports of petroleum products, OECD area, 1966-1970 284
14. Supply and disposal of crude oil, feedstocks and components and petroleum products, OECD area, 1966-1970 285
15. Supply and disposal of crude oil, feedstocks and components and petroleum products, OECD area, 1970 287
16. OECD Member countries' and World tanker tonnage, 1967-1971 288
17. Crude oil exported to the OECD area from certain exporting countries, 1970 290
18. FOB value of oil exports from developing countries and oil exports as a percentage of total exports, 1970 292
19. CIF value of oil imported by OECD countries from countries in process of development and oil imports as a percentage of total imports, 1970 293

Table 1. PRIMARY ENERGY CONSUMPTION BY FORM OF ENERGY, OECD AREA 1960 AND 1970

In percentages of total consumption

	SOLID FUELS		OIL		NATURAL GAS		HYDRO ELECTRICITY		NUCLEAR ENERGY	
	1960	1970	1960	1970	1960	1970	1960	1970	1960	1970
Austria	46.2	24.0	27.8	50.3	14.1	15.4	11.9	10.3	-	-
Belgium	69.2	32.3	30.5	59.3	0.2	8.4	0.1	-	-	-
Denmark	43.5	11.9	56.4	89.5	-	-	0.1	-1.4	-	-
Finland	n.a.	17.5	n.a.	75.5	n.a.	-	n.a.	7.0	-	-
France	56.7	26.5	34.3	62.4	3.3	6.1	5.6	4.1	0.1	0.9
Germany	76.6	39.3	21.7	53.3	0.5	5.8	1.2	1.0	-	0.6
Greece	17.8	22.1	80.6	74.4	-	-	1.7	3.5	-	-
Iceland	4.6	-	82.2	83.3	-	-	13.2	16.7	-	-
Ireland	49.9	18.7	46.0	81.3	-	-	4.1	-	-	-
Italy	17.8	7.6	55.7	77.9	13.7	9.8	12.8	4.2	-	0.5
Luxembourg	92.5	65.9	6.5	31.7	1.0	-	-	2.4	-	-
Netherlands	45.2	8.4	53.5	63.9	1.3	27.6	-	-	-	0.1
Norway	9.8	6.8	45.8	51.7	-	-	44.4	41.5	-	-
Portugal	24.3	13.6	61.9	76.3	-	-	13.8	10.1	-	-
Spain	56.0	26.6	34.4	66.1	-	0.2	9.6	6.6	-	0.5
Sweden	15.3	5.5	66.2	81.2	-	-	18.5	13.3	-	-
Switzerland	23.7	3.8	47.8	75.6	-	-	28.5	17.5	-	3.1
Turkey	65.7	34.8	31.9	62.7	-	-	2.4	2.5	-	-
United Kingdom	73.6	48.0	25.8	44.2	-	5.0	0.2	0.2	0.4	2.6
Total OECD Europe	61.4	29.4	32.5	59.6	1.8	6.7	4.2	3.3	0.1	1.0
Canada	17.5	13.5	55.3	50.7	12.5	23.2	14.7	12.4	-	0.2
USA	23.9	20.5	44.1	42.7	30.3	34.8	1.7	1.7	-	0.3
Total Western Hemisphere	23.4	20.0	44.9	43.3	29.1	33.9	2.6	2.5	-	0.3
Japan	54.8	23.3	36.4	71.7	1.0	1.4	7.8	3.2	-	0.4
Australia	61.7	48.0*	37.0	50.0*	-	0.5*	1.3	1.5*	-	-
Total OECD Area (excluding Australia)	37.4	23.5	40.4	51.2	18.7	21.9	3.4	2.8	0.1	0.6

* 1969 figure.

SOURCE: OECD Energy Statistics, 1972.

265

Table 2. PER CAPITA INLAND CONSUMPTION OF PETROLEUM PRODUCTS OECD AREA 1966-70

Tons per head

	1966	1967	1968	1969	1970
Austria[1]	0.80	0.87	0.99	1.09	1.21
Belgium	1.44	1.58	1.81	2.03	2.30
Denmark	2.22	2.28	2.59	3.15	3.53
Finland	1.64	1.65	1.65	1.94	2.12
France	1.02	1.15	1.27	1.41	1.63
Germany	1.33	1.38	1.53	1.70	1.89
Greece[1]	0.43	0.49	0.54	0.60	0.62
Iceland[1]	2.61	1.56	1.40	1.44	1.47
Ireland	0.84	0.98	0.96	1.13	1.30
Italy	0.86	0.96	1.08	1.24	1.41
Luxembourg	2.84	3.12	3.44	3.78	3.99
Netherlands	1.53	1.58	1.64	1.69	1.84
Norway	1.41	1.44	1.57	1.76	1.90
Portugal[1].............	0.24	0.26	0.29	0.35	0.35
Spain	0.42	0.50	0.60	0.69	0.76
Sweden	2.56	2.48	2.83	3.15	3.51
Switzerland[1]	1.39	1.49	1.63	1.78	1.95
Turkey	0.13	0.16	0.18	0.19	0.21
United Kingdom	1.26	1.35	1.43	1.53	1.63
Average European Area	1.07	1.15	1.26	1.40	1.47
Canada	2.59	2.75	3.00	3.07	3.16
United States	2.87	2.89	3.01	3.19	3.33
Average North America	2.84	2.88	3.01	3.17	3.31
Japan	0.84	1.02	1.19	1.40	1.70
Australia	1.39	1.49	1.62	1.72	1.76
Average Total OECD Area (excluding Australia	1.62	1.70	1.83	1.99	2.10

1. Estimated population.

Table 3.   REFINERY OUTPUT AND CONSUMPTION (INCLUDING BUNKERS)
OF PETROLEUM PRODUCTS OECD AREA 1966 AND 1970
(Million tons)

EUROPEAN AREA

	1966				1970			
	REFINERY OUTPUT[1]	% OF TOTAL	CONSUMP-TION AND BUNKERS[2]	% OF TOTAL	REFINERY OUTPUT[1]	% OF TOTAL	CONSUMP-TION AND BUNKERS[2]	% OF TOTAL
LPG	7.5	2.1	7.9	2.0	10.4	1.8	10.6	1.8
Aviation fuels	9.5	2.6	8.5	2.2	16.6	2.8	14.1	2.4
Motor gasoline	49.9	13.6	53.2	13.8	70.1	12.0	72.5	12.5
Naphtas	-	-	-	-	31.3	5.4	27.1	4.7
Kerosenes	5.9	1.6	5.3	1.4	8.7	1.5	7.7	1.3
Gas/diesel oil	103.5	28.3	112.6	29.3	175.3	30.1	181.6	31.2
Fuel oil	150.8	41.3	159.2	41.4	234.4	40.2	230.3	39.6
Other products	38.3	10.5	38.1	9.9	36.0	6.2	37.6	6.5
Total all products	365.4	100	384.8	100	582.8	100	581.5	100

NORTH AMERICA

	1966				1970			
	REFINERY OUTPUT[1]	% OF TOTAL	CONSUMP-TION AND BUNKERS[2]	% OF TOTAL	REFINERY OUTPUT[1]	% OF TOTAL	CONSUMP-TION AND BUNKERS[2]	% OF TOTAL
LPG	9.7	1.8	33.2	15.3	11.1	1.8	46.8	6.1
Aviation fuels	32.6	6.1	36.2	5.8	41.7	6.6	48.9	6.4
Motor gasoline	219.9	41.1	221.8	35.4	263.2	42.1	267.3	35.2
Naphtas	13.5	2.5	18.2	2.9	5.3	0.8	5.5	0.7
Kerosenes	15.3	2.9	15.5	2.5	14.6	2.3	15.1	2.0
Gas/Diesel oil	119.5	22.4	122.7	19.5	137.3	21.9	143.6	18.9
Fuel oil	47.9	9.0	106.4	17.0	51.6	8.2	137.5	18.1
Other products	76.1	14.2	72.7	11.6	101.6	16.3	95.6	12.6
Total all products	534.5	100	626.7	100	626.4	100	759.9	100

JAPAN

	1966				1970			
	REFINERY OUTPUT[1]	% OF TOTAL	CONSUMP-TION AND BUNKERS[2]	% OF TOTAL	REFINERY OUTPUT[1]	% OF TOTAL	CONSUMP-TION AND BUNKERS[2]	% OF TOTAL
LPG	1.9	2.3	3.6	3.8	3.4	2.1	12.1	6.6
Aviation fuels	1.3	1.6	0.6	0.6	2.0	1.3	1.9	1.0
Motor gasoline	9.2	11.2	8.7	9.2	15.3	9.6	14.9	8.1
Naphtas	-	-	-	-	16.2	10.1	18.7	10.2
Kerosenes	5.1	6.2	4.8	5.1	14.0	8.8	12.0	6.6
Gas/Diesel oil	9.5	11.6	11.3	11.9	19.2	12.0	19.6	10.7
Fuel oil	45.0	54.8	54.2	57.3	82.5	51.7	93.9	51.3
Other products	10.1	12.3	11.4	12.1	7.1	4.4	9.8	5.5
Total all products	82.1	100	94.6	100	159.7	100	182.9	100

AUSTRALIA

	1966				1970			
	REFINERY OUTPUT[1]	% OF TOTAL	CONSUMP-TION AND BUNKERS[2]	% OF TOTAL	REFINERY OUTPUT[1]	% OF TOTAL	CONSUMP-TION AND BUNKERS[2]	% OF TOTAL
LPG	0.1	0.6	0.2	1.2	0.3	1.4	0.3	1.4
Aviation Fuels	0.6	3.7	0.6	3.7	1.1	5.2	1.0	4.5
Motor Gasoline	5.4	33.6	5.9	36.6	7.2	34.1	7.5	33.8
Naphtas	-	-	-	-	-	-	-	-
Kerosenes	0.2	1.2	0.3	1.9	0.2	1.0	0.2	0.9
Gas/Diesel oil	3.1	19.3	2.9	18.0	4.5	21.3	4.4	19.8
Fuel Oil	5.6	34.8	5.0	31.1	6.2	29.4	7.0	31.5
Other Products	1.1	6.0	1.2	7.5	1.6	7.6	1.8	8.1
Total all products	16.1	100	16.1	100	21.1	100	22.2	100

1.   Excluding substitute fuels.
2.   Including substitute fuels.

Table 4.   REFINERY OUTPUT AND CONSUMPTION (INCLUDIN

OECD ARE

MEMBER COUNTRIES	LIQUEFIED PETROLEUM GAS			AVIATION FUELS			MOTOR GASOLINE			KEROSENES		
	REFINERY OUTPUT[1]	CONSUMPTION[2]	SURPLUS/DEFICIT	REFINERY OUTPUT[1]	CONSUMPTION[2]	SURPLUS/DEFICIT	REFINERY OUTPUT[1]	CONSUMPTION[2]	SURPLUS/DEFICIT	REFINERY OUTPUT[1]	CONSUMPTION[2]	SURPLUS/DEFICIT
Austria	0.2	0.2	-	0.1	-	+0.1	1.0	1.6	-0.6	-	-	-
Belgium	0.4	0.5	-0.1	1.1	0.4	+0.7	3.6	2.2	+1.4	0.1	-	+0.1
Denmark	0.2	0.2	-	0.1	0.6	-0.5	1.3	1.5	-0.2	0.1	0.2	-0.1
Finland	-	0.1	-0.1	0.1	0.1	-	1.0	1.0	-	-	-	-
France	2.3	2.1	+0.2	3.0	1.5	+1.5	13.2	12.3	+0.9	-	0.1	-0.1
Germany	2.0	2.2	-0.2	1.5	2.0	-0.5	12.7	15.9	-3.2	0.1	0.1	-
Greece	-	-	-	0.3	0.6	-0.3	0.6	0.6	-	-	0.1	-0.1
Iceland	-	-	-	-	0.1	-0.1	-	-	-	-	-	-
Ireland	-	0.1	-0.1	0.1	0.1	-	0.5	0.6	-0.1	-	0.1	-0.1
Italy	2.1	1.8	+0.3	2.5	1.5	+1.2	12.7	9.3	+3.4	3.3	2.0	+1.3
Luxembourg	-	-	-	-	-	-	-	0.1	-0.1	-	-	-
Netherlands	0.7	0.3	+0.4	2.3	0.7	+1.6	4.6	3.0	+1.6	1.2	1.2	-
Norway	-	-	-	0.2	0.2	-	0.6	0.9	-0.3	0.2	0.4	-0.1
Portugal	-	0.3	-0.3	0.2	0.4	-0.2	0.5	0.5	-	0.2	0.1	+0.1
Spain	1.1	1.4	-0.3	1.3	1.4	-0.1	3.3	2.8	+0.5	0.2	0.2	-
Sweden	-	-	-	0.1	0.5	-0.4	1.3	2.8	-1.5	-	0.2	-0.2
Switzerland	-	-	-	0.1	0.6	-0.5	0.8	2.1	-1.3	-	-	-
Turkey	0.1	0.2	-0.1	-	-	-	1.0	0.9	+0.1	0.4	0.5	-0.1
United Kingdom	1.2	1.2	-	3.5	3.5	-	11.3	14.2	-2.9	2.7	2.5	+0.2
Total OECD Europe	10.3	10.6	-0.3	16.7	14.2	+2.5	69.9	72.2	-2.3	8.5	7.7	+0.8
Canada	0.4	1.5	-1.1	1.9	2.4	-0.5	19.3	19.8	-0.5	2.4	2.7	-0.3
United States	10.7	45.3	-34.6	39.8	46.2	-6.4	243.9	247.5	-3.6	12.2	12.4	-0.2
Total North America	11.1	46.8	-35.7	41.7	48.6	-6.9	263.2	267.3	-4.1	14.6	15.1	-0.5
Japan	3.4	12.1	-8.7	2.0	1.9	+0.1	15.3	14.9	+0.4	14.0	12.0	+2.0
Australia	0.3	0.3	-	1.1	1.0	+0.1	7.2	7.5	-0.3	0.2	0.2	-
Total OECD Area	25.1	69.8	-44.7	61.5	65.7	-4.2	355.6	361.9	-6.3	37.3	35.0	+2.

1. Excluding substitute fuels.
2. Including bunkers and substitute fuels.

Million tons

NAPHTHAS			GAS/DIESEL OIL			FUEL OIL			OTHER PRODUCTS			TOTAL ALL PRODUCTS		
REFINERY OUTPUT[1]	CONSUMPTION[2]	SURPLUS/DEFICIT	REFINERY OUTPUT[1]	CONSUMPTION[2]	SURPLUS/DEFICIT	REFINERY OUTPUT[1]	CONSUMPTION[2]	SURPLUS/DEFICIT	REFINERY OUTPUT[1]	CONSUMPTION[2]	SURPLUS DEFICIT	REFINERY OUTPUT[1]	CONSUMPTION[2]	SURPLUS/DEFICIT
-	0.1	-0.1	1.5	1.6	-0.1	2.6	4.5	-1.9	0.6	1.0	-0.4	6.0	9.0	-3.0
0.1	1.3	-0.2	9.4	8.4	+1.0	11.3	11.1	+0.2	1.2	1.1	+0.1	28.2	25.0	+3.2
0.2	0.1	+0.1	2.9	6.2	-3.3	4.8	8.4	-3.6	0.1	0.6	-0.5	8.7	17.8	-8.1
0.4	0.1	+0.3	2.3	4.0	-1.7	3.5	4.3	-0.6	0.4	0.4	-	7.7	10.0	-2.3
2.2	0.9	+1.3	39.7	37.3	+2.4	27.5	24.3	+3.2	7.7	8.1	-0.4	95.6	86.6	+9.0
4.7	3.7	+1.0	38.2	54.7	-16.5	28.9	29.4	-0.5	10.4	11.9	-1.5	98.5	119.9	-21.4
-	-	-	1.5	1.6	-0.1	1.7	2.5	-0.8	0.3	0.3	-	4.3	5.6	-1.3
-	-	-	-	0.3	-0.3	-	0.1	-0.1	-	-	-	-	0.5	-0.5
-	0.1	-0.1	0.6	0.8	-0.2	1.3	2.0	-0.7	0.1	0.2	-0.1	2.6	4.0	-1.4
10.0	6.5	+3.5	23.0	13.1	+9.9	54.6	45.8	+8.8	3.8	3.8	-	122.2	83.8	+28.4
-	-	-	-	0.5	-0.5	-	0.7	-0.7	-	0.1	-0.1	-	1.4	-1.4
4.2	2.4	+2.8	17.1	8.0	+9.1	25.4	15.5	+9.9	3.0	1.7	+1.3	58.5	32.8	+25.7
0.2	0.2	-	1.6	2.8	-1.2	2.8	2.8	-	-	0.7	-0.7	5.6	8.0	-2.4
0.2	0.2	-	0.7	0.9	-0.2	1.2	1.5	-0.3	0.2	0.3	-0.1	3.2	4.2	-1.0
1.3	0.9	+0.4	7.0	5.2	+1.8	15.6	13.5	+2.1	2.5	1.8	+0.7	32.3	27.2	+5.1
0.1	0.6	-0.5	3.3	9.4	-6.1	5.4	14.8	-9.4	1.2	1.1	+0.1	11.4	29.4	-18.0
0.1	0.2	-0.1	2.2	6.5	-4.3	1.7	2.3	-0.6	0.3	0.6	-0.3	5.2	12.3	-7.1
0.1	0.1	-	1.6	1.9	-0.3	3.3	3.3	-	0.5	0.5	-	7.0	7.4	-0.4
6.4	9.5	-3.1	22.5	17.1	+5.4	42.9	43.3	-0.4	4.2	5.0	-0.8	94.7	96.3	-0.6
31.2	26.9	+4.3	175.1	180.3	-5.2	234.5	230.1	+4.4	36.5	39.2	-2.7	582.9	581.2	+1.5
1.8	1.8	-	17.2	19.2	-2.0	12.9	16.7	-3.8	5.0	5.9	-0.9	60.9	70.0	-9.1
3.5	3.7	-0.2	120.1	124.3	-4.2	38.7	120.7	-82.0	96.6	89.8	+6.8	565.5	689.9	-124.4
5.3	5.5	-0.2	137.3	143.5	-6.2	51.6	137.4	-85.8	101.6	95.7	+5.9	626.4	759.9	-133.5
16.2	18.7	-2.5	19.2	19.5	-0.3	82.5	93.9	-11.4	7.1	9.9	-2.8	159.7	182.9	-23.2
-	-	-	4.5	4.4	+0.1	6.2	7.0	-0.8	1.6	1.8	-0.2	21.1	22.2	-1.1
52.7	51.1	+1.6	336.1	347.7	-11.6	374.8	468.4	-93.6	146.8	146.6	+0.2	1,389.9	1,546.2	-156.3

Million tons

Table 5. ESTIMATED PRINCIPAL END-USES OF OIL PRODUCTS
1966 - 1970

OECD EUROPEAN AREA

YEAR	ROAD	RAIL	AIR	INLAND WATER-WAYS	ELECTRICITY	GAS	IRON STEEL	PETRO-CHEMICAL INDUSTRIES	OTHER INDUSTRIES	OTHER USES	UNALLO-CATED PRODUCTS	TOTAL	BUNKERS	TOTAL ALL PRODUCTS
	1	2	3	4	5	6	7	8	9	10	11	12 = 1 to 11	13	14 = 12 + 13
1966	74.0	3.9	15.0	6.4	26.2	7.8	14.2	-	100.0	77.0	30.4	354.9	29.9	384.8
1967	82.3	4.4	9.9	7.3	28.5	8.4	14.6	14.6	102.2	100.8	26.7	385.1	30.8	415.9
1968	90.8	4.5	11.2	7.7	3.7	9.3	15.4	19.9	113.6	106.6	23.4	434.1	33.4	467.5
1969	89.1	5.8	11.5	5.8	38.9	9.0	17.1	24.0	117.0	117.3	51.0	486.6	38.1	524.7
1970	92.8	3.3	14.1	5.8	49.8	6.8	17.3	26.3	127.7	130.7	69.2	543.8	37.4	581.2

NORTH AMERICA

YEAR	ROAD	RAIL	AIR	INLAND WATER-WAYS	ELECTRICITY	GAS	IRON STEEL	PETRO-CHEMICAL INDUSTRIES	OTHER INDUSTRIES	OTHER USES	UNALLO-CATED PRODUCTS	TOTAL	BUNKERS	TOTAL ALL PRODUCTS
1966	221.8	17.6	36.1	6.8	21.6	-	5.5	-	11.4	241.2	53.8	615.8	11.0	626.0
1967	228.9	12.3	42.8	5.5	24.3	0.2	4.5	20.1	20.4	227.1	44.9	631.0	11.7	642.7
1968	243.6	13.1	48.6	4.7	28.2	0.3	4.3	26.3	21.1	155.7	120.5	666.4	12.3	678.7
1969	255.9	12.1	51.0	2.0	44.2	0.8	5.6	30.0	31.3	169.5	108.6	910.4	18.3	728.7
1970	295.2	18.8	48.5	-	54.4	0.5	6.0	30.3	33.1	222.3	40.2	749.3	10.6	759.9

JAPAN

YEAR	ROAD	RAIL	AIR	INLAND WATER-WAYS	ELECTRICITY	GAS	IRON STEEL	PETRO-CHEMICAL INDUSTRIES	OTHER INDUSTRIES	OTHER USES	UNALLO-CATED PRODUCTS	TOTAL	BUNKERS	TOTAL ALL PRODUCTS
1966	8.9	0.9	0.6	1.0	12.2	0.7	-	-	30.6	12.8	15.6	83.3	11.3	94.6
1967	14.4	1.2	0.6	1.9	16.7	0.8	-	7.3	32.8	8.4	18.5	102.6	9.7	112.3
1968	15.9	1.2	0.7	2.6	23.2	-	-	9.2	41.3	18.4	7.6	120.1	7.4	127.5
1969	18.6	1.3	0.7	3.0	28.9	-	0.5	14.9	49.1	16.2	10.1	143.3	9.6	152.9
1970	27.0	0.5	1.9	3.6	31.8	1.5	11.0	17.3	54.5	21.0	16.1	176.2	6.7	182.9

Table 6. CRUDE OIL PRODUCTION OECD AREA 1966-1970

EUROPEAN AREA

Thousand tons

PRODUCING COUNTRIES	1966	1967	1968	1969	1970
Austria	2,757.1	2,684.9	2,724.3	2,758.2	2,798.2
France	3,608.0	2,832.0	2,688.0	2,499.0	2,309.0
Germany	7,880.0[1]	7,927.0	7,982.0	7,876.0	7,535.0
Italy	1,849.7	1,616.0	1,506.0	1,479.4	3,659.5[2]
Netherlands	2,366.0	2,265.0	2,147.0	2,020.0	1,919.0
Spain	-	84.0	127.0	140	-
Sweden	25.0	-	-	-	-
Turkey	2,041.1	2,752.0	3,104.0	3,599.2	3,542.0
United Kingdom	78.0	89.0	81.0	77.0	83.0
Total	20,604.9	20,249.9	20,359.3	20,448.8	21,845.7

1. Including 12,000 metric tons of natural gas liquids.
2. Of which 2,254,000 tons of feedstocks.

NORTH AMERICA

PRODUCING COUNTRIES	1966	1967	1968	1969	1970
Canada	46,519.0	47,286.0	51,075.0	55,266.0	62,047.0
United States	449,393.2	433,600.8	448,760.8	454,536.0	474,309.0
Total	495,912.2	480,886.8	499,835.8	509,802.0	536,356.0

AUSTRALIA

	1966	1967	1968	1969	1970
Australia	430.7	992.2	1,828.9	2,081.3	8,147.4

Table 7. SOURCES OF IMPORTS OF CRUDE, FEEDSTOCKS AND COMPONENTS AND PETROLEUM PRODUCTS 1966-1970

OECD EUROPEAN AREA

Million tons

a) Crude and Process Oils

	OECD AREA		WESTERN HEMISPHERE		WESTERN EUROPE OUTSIDE OECD + YUGOSLAVIA	AFRICA				MIDDLE EAST					FAR EAST	USSR AND EAST EUROPE	OTHERS UNALLOCATED	TOTAL OUTSIDE OECD EUROPEAN AREA
	EUROPE	USA CANADA	VENEZUELA	OTHERS		ALGERIA	LIBYA	WEST AFRICA	OTHERS	KUWAIT	SAUDI ARABIA	IRAN	IRAQ	OTHERS				
YEAR	1	2	3	4	5	6	7	8	9	10	11	12	13	14	15	16	17	18=2 to 17
1966	1.5	-	21.0	2.8	0.4	29.0	64.3	12.4	6.2	65.3	56.8	29.5	47.1	19.2	2.0	15.2	0.1	371.3
1967	1.8	2.8	25.8	3.0	0.4	33.7	75.4	14.0	3.8	62.1	60.0	38.6	43.0	23.6	1.6	18.9	-	406.7
1968	2.3	0.1	24.9	2.0	0.3	39.0	113.6	4.0	2.9	67.0	61.9	37.4	56.8	35.8	0.8	25.2	0.3	472.0
1969	3.5	0.3	25.4	1.1	0.4	40.4	136.0	19.4	2.7	73.9	68.6	41.2	56.3	42.4	0.5	22.4	-	531.0
1970	4.1	0.1	22.8	1.0	0.2	43.1	150.7	35.9	6.7	79.5	82.4	44.2	52.9	49.7	0.4	23.7	0.4	593.7

b) Petroleum products

	OECD AREA		WESTERN HEMISPHERE			AFRICA	NEAR AND MIDDLE EAST	FAR EAST	EASTERN EUROPE			WESTERN EUROPE OUTSIDE OECD	UNALLOCATED	TOTAL IMPORTS FROM OUTSIDE OECD EUROPEAN AREA
	EUROPE	USA CANADA	VENEZUELA	NETHERLANDS	OTHERS				USSR	RUMANIA	OTHERS			
YEAR	1	2	3	4	5	6	7	8	9	10	11	12	13	14=2 to 13
1966	67.7	2.3	5.2	4.1	4.7	1.0	8.7	0.6	7.5	1.9	1.7	0.8	4.0	42.5
1967	70.8	2.9	4.2	2.7	3.4	1.1	6.9	0.5	9.2	1.7	2.8	1.0	2.3	38.7
1968	75.0	2.2	2.8	2.3	3.7	1.2	4.8	0.2	13.2	2.5	2.7	1.1	2.9	39.6
1969	84.4	3.2	2.4	1.5	2.7	1.1	4.8	0.2	12.4	2.0	3.3	0.7	0.9	35.2
1970	97.9	3.2	1.2	1.4	3.0	0.8	4.5	-	14.6	1.8	2.9	0.7	1.5	36.1

272

Table 7. (Cont'd)

AUSTRALIA

a) Crude and Process Oils

YEAR	OECD AREA		WESTERN HEMISPHERE		WESTERN EUROPE OUTSIDE OECD + YUGOSLAVIA	AFRICA				MIDDLE EAST					FAR-EAST	USSR AND EAST EUROPE	OTHERS UNALLOCATED	TOTAL OUTSIDE OECD EUROPEAN AREA
	EUROPE	USA AND CANADA	VENEZUELA	OTHERS		ALGERIA	LIBYA	WEST AFRICA	OTHERS	KUWAIT	SAUDI ARABIA	IRAN	IRAQ	OTHERS				18=2 to 17
	1	2	3	4	5	6	7	8	9	10	11	12	13	14	15	16	17	
1966									0.1	3.3	3.9	1.5	1.3	2.4	5.4			17.9
1967										4.1	4.5	1.6	1.3	2.3	5.7			19.5
1968										5.1	3.9	1.4	1.1	2.1	6.4			20.0
1969										4.0	4.0	0.6	1.0	4.5	6.1			20.2
1970										4.3	3.2	0.4	1.4	3.2	3.4			15.9

b) Petroleum products

YEAR	OECD AREA		WESTERN HEMISPHERE			AFRICA	NEAR AND MIDDLE EAST	FAR EAST	EASTERN EUROPE			WESTERN EUROPE OUTSIDE OECD	UNALLOCATED	TOTAL IMPORTS FROM OUTSIDE OECD EUROPEAN AREA
	EUROPE	USA AND CANADA	VENEZUELA	NETHERLANDS	OTHERS				USSR	RUMANIA	OTHERS			14 = 2 to 13
	1	2	3	4	5	6	7	8	9	10	11	12	13	
1966		0.1					0.7	0.3						1.1
1967		0.1					0.5	0.3						0.9
1968		0.1					0.4	0.4						0.9
1969		0.2					1.6	0.4					0.1	2.3
1970		0.2					1.7	0.6					0.1	2.6

273

Table 7. (Cont'd)
NORTH AMERICA

Million tons

a) Crude oil, feedstocks and components

| YEAR | OECD AREA | | WESTERN HEMISPHERE | | | AFRICA | | | | MIDDLE EAST | | | | | FAR EAST | OTHER EUROPE USSR | UNALLOCATED | TOTAL IMPORTS FROM OUTSIDE NORTH AMERICAN AREA |
| | CANADA USA | EUROPE | VENEZUELA | OTHERS | WEST EUROPE YUGOSLAVIA, ETC. | ALGERIA | LIBYA | WEST AFRICA | OTHERS | KUWAIT | SAUDI ARABIA | IRAN | IRAQ | OTHERS | | | | |
	1	2	3	4	5	6	7	8	9	10	11	12	13	14	15	16	17	18=2 to 17
1966	17.4	0.1	33.1	4.8	–	0.2	4.0	0.6	1.7	1.9	8.4	7.1	2.3	2.7	2.7	–	–	69.6
1967	21.0	0.1	35.4	5.2	–	0.2	2.7	0.3	1.2	1.1	5.9	5.6	0.7	1.5	3.3	–	–	63.2
1968	22.9	0.1	36.4	6.3	–	0.2	5.5	0.8	–	2.6	5.1	5.7	–	4.6	3.6	–	–	70.9
1969	27.4	–	34.4	7.6	–	–	7.0	3.5	–	2.3	3.8	4.6	0.6	6.6	4.4	–	–	74.8
1970	33.0	0.1	33.8	5.8	–	0.3	2.3	4.8	–	2.6	4.1	4.5	1.4	4.7	3.5	–	–	67.9

b) Petroleum products

| YEAR | OECD AREA | | WESTERN HEMISPHERE | | | AFRICA | NEAR AND MIDDLE EAST | FAR EAST | EASTERN EUROPE | | | WEST EUROPE OUTSIDE OECD AND YUGOSLAVIA | UNALLOCATED | TOTAL IMPORTS FROM OUTSIDE NORTH AMERICAN AREA |
| | CANADA USA | EUROPE | VENEZUELA | NETHERLANDS ANTILLES | OTHERS | | | | USSR | RUMANIA | OTHERS | | | |
	1	2	3	4	5	6	7	8	9	10	11	12	13	14=2 to 13
1966	3.0	3.2	36.3	21.0	15.3	–	1.4	0.4	–	–	–	–	0.1	77.7
1967	3.3	3.2	32.4	23.1	16.6	0.4	0.8	–	–	–	–	–	–	76.5
1968	3.6	6.5	33.3	23.4	19.9	0.2	1.1	–	0.1	–	–	–	–	84.4
1969	4.1	8.9	34.2	25.7	22.5	0.3	1.0	–	0.1	0.3	–	–	–	93.1
1970	5.3	9.2	41.1	28.1	29.1	0.3	0.5	–	0.2	0.4	–	–	–	108.9

274

Table 7. (Cont'd)

JAPAN

a) Crude oil, feedstocks and components

Million tons

YEAR	OECD AREA			AFRICA	MIDDLE EAST						FAR EAST		UNALLOCATED	TOTAL IMPORTS FROM OUTSIDE JAPAN
	OECD EUROPEAN MEMBERS	WESTERN HEMISPHERE MEMBERS	WESTERN HEMISPHERE		KUWAIT	SAUDI ARABIA	IRAN	IRAQ	ABU-DHABI	OTHERS	INDONESIA	OTHERS		
	1	2	3	4	5	6	7	8	9	10	11	12	13	14=1 to 13
1966	-	0.1	0.4	-	17.5	14.7	23.6	4.6	0.5	14.6	5.0	0.3	3.1	84.4
1967	-	0.1	0.4	-	18.4	18.8	36.4	2.8	2.3	17.8	6.0	0.7	0.1	103.3
1968	-	0.1	0.5	-	16.4	22.7	45.3	1.7	6.0	18.6	9.0	0.4	0.8	121.5
1969	-	0.1	0.5	0.1	12.8	25.5	61.0	0.2	6.5	22.0	14.0	1.0	0.4	144.1
1970	-	-	0.6	1.0	23.2	33.3	73.8	-	7.8	6.7	22.0	0.6	0.5	169.5

b) Petroleum products

YEAR	OECD AREA			AFRICA	EASTERN EUROPE	MIDDLE EAST				FAR EAST	OTHERS	TOTAL IMPORTS FROM OUTSIDE JAPAN
	OECD EUROPEAN MEMBERS	WESTERN HEMISPHERE	WESTERN HEMISPHERE			KUWAIT	SAUDI ARABIA	IRAN	IRAQ			
	1	2	3	4	5	6	7	8	9	10	11	12=1 to 11
1966	-	1.9	1.5	-	1.8	1.2	1.3	1.8	1.4	2.3	-	13.2
1967	-	2.9	1.8	-	1.8	1.0	2.1	1.2	1.1	2.7	0.2	14.8
1968	-	3.3	2.2	-	1.6	1.7	2.0	1.1	1.2	4.4	0.2	17.7
1969	-	*3.0	1.0	-	0.8	2.8	2.3	1.8	0.9	6.1	-	18.7
1970	0.2	4.3	0.8	-	1.3	4.4	3.1	2.6	2.3	8.1	0.2	27.3

275

Table 8. SOURCES OF IMPORTS OF CRUDE OIL,
OECD

COUNTRIES	IMPORTED								
	OECD AREA				WESTERN HEMISPHERE				WESTERN EUROPE
	NORTH AMERICA	EUROPE	JAPAN	TOTAL	VENEZUELA	NETHERLANDS ANTILLES	COLUMBIA	OTHERS	
	1	2	3	4-1-3	5	6	7	8	9
Austria	.	0.17		0.17	-	-	-	-	0.12
Belgium		0.03	-	0.03	2.85	-	-	-	-
Denmark	-	0.20	-	0.20	0.37	-	-	-	-
Finland	-	-	-	-	0.26	-	-	-	-
France	-	-	-	-	2.45	-	-	-	-
Germany	0.02	1.39	0.01	1.42	3.40	0.10	-	-	0.10
Greece	-	-	-	-	-	-	-	-	-
Iceland	-	-	-	-	-	-	-	-	-
Ireland	-	0.05	-	0.05	-	-	-	-	-
Italy	-	-	-	-	2.33	-	-	-	-
Luxembourg	-	-	-	-	-	-	-	-	-
Netherlands	-	0.49	-	0.49	1.19	0.27	-	-	-
Norway	-	-	-	-	1.22	-	-	-	-
Portugal	-	-	-	-	-	-	-	-	-
Spain	-	-	-	-	2.13	-	-	-	-
Sweden	0.03	-	-	0.03	1.81	-	-	-	-
Switzerland	-	0.16	-	0.16	-	-	-	-	-
Turkey	-	-	-	-	-	-	-	-	-
United Kingdom	0.04	1.64	-	1.68	4.81	0.46	0.03	0.28	-
Total OECD Europe	0.09	4.13	0.01	4.23	22.82	0.	0.03	0.28	0.22
Canada	-	-	-	-	18.58	-	1.00	0.04	-
USA	33.02	-	0.09	33.11	6.34	0.83	1.10	2.88	-
Total North America	33.02	-	0.09	33.11	24.92	0.83	2.10	9.92	-
Japan	0.06	-	-	0.06	0.56	-	-	-	-
Australia	-	-	-	-	-	-	-	-	-
OECD Area Total	33.17	4.13	0.10	37.40	48.3	1.57	2.13	3.20	0.22

276

Million tons

ROM :

EASTERN EUROPE	AFRICA				MIDDLE EAST								FAR EAST	TOTAL IMPORTS FROM OUTSIDE OECD AREA	TOTAL IMPORTS
	ALGERIA	LIBYA	NIGERIA	OTHERS	KUWAIT	SAUDI ARABIA	IRAQ	IRAN	QUATAR	ABU DHABI	EGYPT	OTHERS			
10	11	12	13	14	15	16	17	18	19	20	21	22	23	24=5-23	25=4+24
2.11	-	0.42	-	0.08	-	0.07	0.45	-	-	-	-	0.14	-	3.39	3.56
0.54	1.54	6.09	0.73	0.23	5.24	5.08	1.09	3.50	1.11	1.25	0.23	0.31	0.07	29.86	29.89
-	0.11	1.28	0.77	1.05	2.84	1.41	-	0.65	-	-	0.21	1.28	-	9.97	10.17
6.70	-	0.03	-	-	-	-	-	2.76	-	-	-	-	-	9.75	9.75
1.42	26.99	17.64	5.19	2.12	11.09	9.45	12.16	3.79	1.76	4.52	0.62	2.14	-	101.34	101.34
3.45	7.98	40.92	6.94	0.99	3.95	12.06	3.48	8.27	0.31	5.76	1.27	-	-	98.89	100.31
-	-	-	-	-	-	-	-	-	-	-	-	-	-	4.6	4.6
-	-	-	-	-	-	-	-	-	-	-	-	-	-	-	-
-	-	0.04	-	-	0.86	0.84	0.24	0.65	-	-	-	0.05	-	2.68	2.73
8.78	2.84	35.72	0.54	-	13.82	16.28	21.58	6.36	1.38	0.08	1.52	2.89	-	114.12	114.12
-	-	-	-	-	-	-	-	-	-	-	-	-	-	-	-
0.03	0.25	12.17	7.72	1.33	11.63	9.23	5.04	6.08	1.24	0.12	0.38	2.13	0.25	59.06	59.55
-	0.05	0.60	1.11	-	0.68	1.66	-	0.22	0.14	-	-	1.44	-	7.12	7.12
-	-	-	-	0.11	-	0.64	2.03	0.12	-	0.79	-	-	-	3.69	3.69
-	1.24	8.01	2.2	0.15	2.07	8.53	1.38	2.26	-	-	1.31	2.85	-	32.17	32.17
0.69	-	0.39	3.1	-	0.62	-	0.16	1.09	0.79	1.04	-	2.03	-	11.72	11.75
-	0.75	3.21	-	0.16	0.61	0.62	-	-	-	-	-	-	-	5.35	5.51
-	0.03	0.49	-	-	0.40	2.87	-	-	-	-	-	0.05	-	3.84	3.84
-	1.32	23.73	7.61	0.51	26.11	16.08	2.39	8.43	4.18	-	0.32	3.97	0.07	100.3	101.98
23.72	43.10	150.74	35.91	6.73	79.52	82.35	52.87	44.18	10.91	13.56	5.86	19.32	0.39	597.85	602.08
-	-	-	2.46	-	0.73	1.86	1.21	2.60	-	-	-	0.67	-	29.15	-
-	0.27	2.26	2.38	-	1.67	2.28	-	1.65	-	3.08	1.11	-	3.49	29.34	62.45
-	0.27	2.26	4.84	-	2.40	4.14	1.21	4.25	-	3.08	1.11	0.67	3.49	58.49	91.60
0.5	-	0.33	-	0.63	23.25	33.31	-	73.75	0.18	7.81	1.28	5.22	22.58	169.4	169.40
-	-	-	-	-	4.3	3.2	1.4	0.4	1.3	-	-	1.9	3.4	15.9	15.9
24.22	43.37	-	40.75	7.36	109.47	123.0	55.48	122.58	12.39	24.45	8.25	27.11	29.86	841.64	879.04

Table 9.   SOURCES OF IMPORTS OF PETROLEUM PRODUCTS BY COUNTRIES

OECD AREA 1970

Million tons

IMPORTED FROM:

COUNTRIES	NORTH AMERICA (1)	OECD AREA — EUROPE (2)	OECD AREA — JAPAN (3)	OECD AREA — TOTAL (4)	WESTERN HEMISPHERE — VENEZUELA (5)	WESTERN HEMISPHERE — NETHERL. ANTILLES (6)	WESTERN HEMISPHERE — TRINIDAD (7)	WESTERN HEMISPHERE — OTHERS (8)	WESTERN EUROPE (9)	EASTERN EUROPE — USSR (10)	EASTERN EUROPE — RUMANIA (11)	EASTERN EUROPE — OTHERS (12)	AFRICA (13)	MIDDLE EAST (14)	FAR EAST (15)	UNALLOCATED (16)	TOTAL OUTSIDE OECD (17=5-16)	TOTAL IMPORTS (18=4+17)
Austria	–	2.6	–	2.6	–	–	–	–	0.1	–	0.1	0.5	–	–	–	–	0.7	3.3
Belgium	0.3	4.8	–	5.1	0.1	0.1	–	–	0.1	0.4	–	0.2	–	0.3	–	–	1.2	6.3
Denmark	–	9.3	–	9.3	–	0.1	0.3	–	–	0.6	0.2	0.2	–	0.3	–	–	1.5	10.8
Finland	–	0.2	–	0.2	0.1	–	–	–	–	2.7	–	0.1	–	0.1	–	–	3.0	3.2
France	0.2	4.4	–	4.6	–	–	–	–	–	1.1	0.4	–	0.1	0.1	–	–	1.8	6.4
Germany	1.3	24.4	–	25.7	–	0.1	0.1	–	0.2	2.4	0.6	1.3	–	0.6	–	–	5.3	31.0
Greece	–	–	–	–	–	–	–	–	–	–	–	–	–	–	–	–	–	(1.6)
Iceland	–	0.1	–	0.1	–	–	–	0.1	–	0.4	–	–	–	–	–	–	0.5	0.6
Ireland	–	2.0	–	2.0	–	–	–	–	–	0.2	–	–	–	–	–	–	0.3	2.3
Italy	0.4	0.6	–	1.0	–	0.1	–	–	0.2	1.0	0.2	0.1	0.1	0.1	–	0.5	2.2	3.2
Luxembourg	–	1.4	–	1.4	–	–	–	–	–	–	–	–	–	–	–	–	–	1.4
Netherlands	0.2	6.6	–	6.8	0.2	0.1	0.2	0.2	–	0.5	–	–	0.1	0.7	–	1.0	2.8	9.6
Norway	0.3	2.9	–	3.2	0.1	0.1	–	–	–	0.4	–	–	–	0.2	–	–	1.8	4.2
Portugal	–	1.1	–	1.1	–	0.1	–	–	–	–	–	–	0.1	0.1	–	–	0.3	1.2
Spain	0.2	0.8	–	1.0	–	–	–	–	–	–	–	–	0.1	–	–	–	0.2	1.2
Sweden	0.1	13.5	–	13.6	0.5	0.1	1.3	0.1	–	4.2	0.1	0.3	0.1	0.2	–	–	6.9	20.5
Switzerland	–	7.2	–	7.2	–	–	–	–	–	0.3	–	0.2	–	–	–	–	0.5	7.7
Turkey	–	0.3	–	0.3	–	–	–	–	–	0.2	0.1	–	–	–	–	–	0.3	0.6
United Kingdom	0.2	15.9	–	16.1	0.6	0.6	0.4	0.4	0.1	–	0.1	–	0.2	1.8	–	–	4.2	20.3
Total OECD Europe	3.2	98.1	–	101.3	1.6	1.4	2.3	0.8	0.7	14.5	1.8	2.9	0.8	4.4	–	1.5	32.7	135.6
Canada	1.7	0.3	–	2.0	3.8	3.5	0.2	0.5	–	–	–	–	–	–	–	–	8.0	10.0
United States	3.6	9.0	–	12.6	37.3	24.6	11.4	17.0	–	0.2	0.4	–	0.3	0.5	–	–	91.7	104.3
Total North America	5.3	9.3	–	14.6	41.1	28.1	11.6	17.5	–	0.2	0.4	–	0.3	0.5	–	–	99.7	114.3
Japan	4.3	0.2	–	4.5	0.2	0.5	–	0.1	–	1.2	–	–	–	12.4	8.1	0.1	22.6	27.1
Australia	0.2	–	–	0.2	–	–	–	–	–	–	–	–	–	1.7	0.6	0.1	2.4	2.6
Total OECD Area	13.0	107.6	–	120.6	42.9	30.0	13.9	18.4	0.7	15.9	2.2	2.9	1.1	19.0	8.7	1.7	159.0	279.6

278

Table 10. EXPORTS OF PETROLEUM PRODUCTS WITH DESTINATION AND ANNUAL TONNAGE OF BUNKERS 1966 - 1970

Million tons

EUROPEAN AREA - OECD AREA

| | OECD AREA | | | WESTERN EUROPE OUTSIDE OECD AND YUGOSLAVIA | USSR AND EASTERN EUROPE | AFRICA | FAR AND MIDDLE EAST | UNALLOCATED | TOTAL EXPORTS OUTSIDE OECD EUROPE AREA | TOTAL BUNKERS |
	EUROPEAN MEMBER COUNTRIES	UNITED STATES AND CANADA	WESTERN HEMISPHERE							
	1	2	3	4	5	6	7	8	9 = 2 TO 8	10
1966	71.9	1.9	0.2	0.6	0.1	2.5	2.0	0.3	7.6	29.9
1967	73.5	2.7	0.3	1.6	0.2	3.0	1.3	0.3	9.4	30.9
1968	79.1	6.5	0.5	1.6	0.3	2.8	1.0	2.2	14.9	33.4
1969	89.4	9.6	1.6	0.7	0.2	3.5	1.8	1.3	18.7	38.1
1970	103.2	9.4	1.3	0.8	0.4	3.6	1.2	1.6	18.3	37.4

AUSTRALIA

| | OECD AREA | | | WESTERN HEMISPHERE | OTHER WESTERN EUROPEAN | AFRICA | FAR AND MIDDLE EAST | USSR AND EUROPE | UNALLOCATED | TOTAL EXPORTS AUSTRALIA | TOTAL BUNKERS |
	UNITED STATES AND CANADA	EUROPEAN MEMBER COUNTRIES	JAPAN								
	1	2	3	4	5	6	7	8	9	10	11
1966		0.1	0.1			0.1	0.3		0.5	1.1	1.8
1967		0.1	0.1			-	0.3		0.6	1.1	2.0
1968		-	0.2			0.1	0.2		0.6	1.1	2.3
1969		-	-			-	0.3		0.5	0.8	2.2
1970		-	0.1			0.1	0.7		0.5	1.4	2.3

279

Million tons

Table 10. (Contd.)
NORTH AMERICA

	OECD AREA			WESTERN HEMISPHERE	OTHER WESTERN EUROPEAN	AFRICA	MIDDLE/NEAR FAR/EAST	USSR AND EAST EUROPE	UNALLOCATED	TOTAL EXPORTS OUTSIDE NORTH AMERICA	TOTAL BUNKERS
	UNITED STATES AND CANADA	EUROPEAN MEMBER COUNTRIES	JAPAN								
	1	2	3	4	5	6	7	8	9	10 = 2 - 9	11
1966	2.3	1.7	1.6	1.6	-	-	0.1	-	3.5	6.9	11.0
1967	3.6	2.9	3.0	2.5	-	0.2	0.7	-	0.5	6.8	11.7
1968	3.9	3.0	3.3	2.0	-	0.3	0.7	-	0.4	6.4	12.3
1969	4.5	3.2	2.9	2.5	-	0.3	0.7	-	0.6	7.3	18.3
1970	5.0	3.9	4.3	2.5	-	0.2	0.6	-	0.6	12.1	10.6

JAPAN

	OECD AREA		WESTERN HEMISPHERE	OTHER WESTERN EUROPEAN	AFRICA	MIDDLE EAST	FAR EAST	USSR AND EAST EUROPE	UNALLOCATED	TOTAL EXPORTS OUTSIDE JAPAN	TOTAL BUNKERS
	UNITED STATES AND CANADA	EUROPEAN MEMBER COUNTRIES									
	1	2	3	4	5	6	7	8	9	10	11
1966	0.1	-	-	-	-	-	0.9	-	0.7	1.6	11.3
1967	-	-	-	-	-	-	1.0	-	1.4	2.4	9.7
1968	-	-	-	-	-	-	0.6	-	1.8	2.4	7.4
1969	0.2	-	-	-	-	-	0.8	-	11.0	11.8	9.6
1970	0.2	-	-	-	-	-	0.6	-	8.7	9.3	6.7

Million tons

Table 11. OECD AREA EXPORTS OF PETROLEUM PRODUCTS BY DESTINATION - 1970

EXPORTING COUNTRIES	OECD AREA				TOTAL EXPORTS TO:							TOTAL EXPORTS TO OUTSIDE OECD AREA	GRAND TOTAL
	USA CANADA	EUROPE	JAPAN	OECD TOTAL	WESTERN HEMISPHERE	WESTERN EUROPE	EASTERN EUROPE	AFRICA	NEAR AND MIDDLE EAST	FAR EAST	UNALLOCATED	12 = 5 - 11	13 = 12 + 4
	1	2	3	4	5	6	7	8	9	10	11	12	13
Austria	-	-	-	-	-	-	0.1	-	-	-	0.1	0.2	0.2
Belgium	0.3	8.7	-	9.0	-	-	-	-	-	-	-	-	9.0
Denmark	-	1.9	-	1.9	-	-	-	-	-	-	-	-	1.9
Finland	-	0.5	-	0.5	-	-	-	-	-	-	-	-	0.5
France	0.2	10.0	-	10.2	-	-	0.1	0.4	-	-	-	0.5	10.7
Germany	-	8.5	-	8.5	-	-	0.1	-	-	-	-	0.1	8.6
Greece	-	0.1	-	0.1	-	0.1	-	0.1	-	-	-	0.2	0.3
Iceland	-	-	-	-	-	-	-	-	-	-	-	-	-
Ireland	-	0.6	-	0.6	-	-	-	-	-	-	0.2	0.2	0.8
Italy	4.2	20.9	-	25.1	0.4	0.5	-	1.1	0.8	-	0.9	3.6	28.7
Luxembourg	-	-	-	-	-	-	-	-	-	-	-	-	-
Netherlands	2.5	28.8	-	31.4	0.4	0.2	-	1.5	0.2	-	-	2.3	33.7
Norway	-	1.6	-	1.6	-	-	-	-	-	-	-	-	1.6
Portugal	-	0.4	-	0.4	-	-	-	0.1	-	-	-	0.1	0.5
Spain	1.5	3.2	-	4.7	0.3	-	-	0.1	-	-	0.3	0.7	5.4
Sweden	-	1.8	-	1.8	-	-	-	-	-	-	-	-	1.8
Switzerland	-	0.3	-	0.3	-	-	-	-	-	-	-	-	0.3
Turkey	-	-	-	-	-	-	-	-	-	-	-	-	-
United Kingdom	0.6	16.0	-	16.6	0.2	-	-	0.4	-	-	-	0.6	17.2
Total OECD Europe	9.3	103.3	0.1	112.7	1.3	0.8	0.3	3.7	1.0	-	1.4	8.5	121.2
Canada	3.3	-	0.3	3.6	0.1	-	-	-	-	-	-	0.1	3.7
United States	1.7	3.9	4.1	9.7	2.5	-	-	0.2	-	0.6	0.6	3.9	13.6
Total North America	5.0	3.9	4.4	13.3	2.6	-	-	0.2	-	0.6	0.6	3.9	17.3
Japan	0.2	-	-	0.2	-	-	-	-	-	0.6	8.7	9.3	9.5
Australia	-	-	0.1	0.1	-	-	-	0.1	0.7	-	0.5	1.3	1.4
Total OECD Area	14.5	107.2	4.6	126.3	3.9	0.8	0.3	4.0	1.7	1.2	11.2	23.1	149.4

Table 12. INTRA-OECD AREA TRADE

IMPORTED INTO:	AUSTRALIA	AUSTRIA	BELGIUM	DENMARK	FINLAND	FRANCE	GERMANY	GREECE	ICELAND	IRELAND
Australia	-	-	-	-	-	-	-	-	-	-
Austria	-	-	-	-	-	-	1.3	-	-	-
Belgium	-	-	-	-	-	0.3	0.3	-	-	-
Denmark	-	-	0.4	-	-	-	0.5	-	-	-
Finland	-	-	-	0.1	-	0.2	-	-	-	-
France	-	-	0.1	-	-	-	1.1	-	-	-
Germany	-	-	2.5	0.1	-	2.8	-	-	-	-
Greece	-	-	-	-	-	-	-	-	-	-
Iceland	-	-	-	-	-	-	-	-	-	-
Ireland	-	-	-	-	-	0.1	0.1	-	-	-
Italy	-	-	-	-	-	0.1	0.1	-	-	-
Luxembourg	-	-	0.8	-	-	0.1	0.3	-	-	-
Netherlands	-	-	1.1	-	0.1	0.9	1.4	-	-	-
Norway	-	-	0.2	0.3	-	0.1	-	-	-	-
Portugal	-	-	-	-	-	0.1	-	-	-	-
Spain	-	-	-	-	-	0.3	-	-	-	-
Sweden	-	-	1.4	1.4	0.1	.0.1	0.2	-	-	-
Switzerland	-	-	0.4	-	-	2.0	2.3	-	-	-
Turkey	-	-	-	-	-	-	-	-	-	-
United Kingdom	-	-	1.1	-	0.3	2.1	0.7	-	-	0.6
Total Europe	-	-	8.0	1.9	0.5	9.2	8.3	-	-	0.6
Canada	-	-	-	-	-	-	-	-	-	-
United States	-	-	0.3	-	-	0.2	-	-	-	-
Total North America	-	-	0.3	-	-	0.2	-	-	-	-
Japan	0.1	-	-	-	-	-	-	-	-	-
Total OECD	0.1	-	8.3	1.9	0.5	9.4	8.3	-	-	0.6

Million tons

EXPORTED FROM:

ITALY	LUXEMBOURG	NETHERLANDS	NORWAY	PORTUGAL	SPAIN	SWEDEN	SWITZERLAND	TURKEY	UNITED KINGDOM	TOTAL OECD EUROPE	CANADA	UNITED STATES	TOTAL NORTH AMERICA	JAPAN	TOTAL OECD AREA
-	-	-	-	-	-	-	-	-	-	-	-	-	-	-	-
0.9	-	-	-	-	-	-	0.3	-	-	2.5	-	-	-	-	2.5
0.5	-	3.5	-	-	-	-	-	-	0.2	4.8	-	0.3	0.3	-	5.1
0.7	-	1.4	0.1	-	0.3	0.9	-	-	4.9	9.2	-	-	-	-	9.2
-	-	-	-	-	-	-	-	-	0.1	0.4	-	-	-	-	0.4
2.7	-	0.2	-	-	0.1	-	0.1	-	0.1	4.4	-	0.2	0.2	-	4.6
3.1	-	14.3	0.1	-	0.5	-	0.1	-	0.9	24.4	-	1.3	1.3	-	25.7
-	-	-	-	-	-	-	-	-	-	-	-	-	-	-	-
-	-	-	-	-	-	-	-	-	0.1	0.1	-	-	-	-	0.1
-	-	0.1	-	-	-	-	-	-	1.7	2.0	-	-	-	-	2.0
-	-	-	-	-	0.2	-	-	-	-	0.4	-	0.4	0.4	-	0.8
-	-	0.1	-	-	-	-	-	-	-	1.3	-	-	-	-	1.3
1.6	-	-	-	0.1	0.2	-	-	-	1.0	6.4	-	0.2	0.2	-	6.6
-	-	0.4	-	-	0.1	0.3	-	-	1.6	3.0	-	0.3	0.3	-	3.3
0.3	-	0.3	-	-	0.2	-	-	-	0.2	1.1	-	-	-	-	1.1
0.2	-	0.2	-	-	-	-	-	-	0.1	0.8	-	0.2	0.2	-	1.0
1.2	-	2.2	1.1	-	0.4	-	-	-	5.4	13.5	-	0.1	0.1	-	13.6
1.7	-	0.8	-	-	-	-	-	-	-	7.2	-	-	-	-	7.2
0.1	-	-	-	-	-	-	-	-	0.1	0.2	-	-	-	-	0.2
3.4	-	6.5	0.2	0.2	0.7	0.2	-	-	-	16.0	-	0.2	0.2	-	16.2
16.4	-	30.0	1.5	0.3	2.7	1.4	0.5	-	16.4	97.7	-	3.2	3.2	-	100.9
0.1	-	-	-	-	0.1	-	-	-	-	0.2	-	1.7	1.7	-	1.9
4.5	-	2.1	-	-	1.2	-	-	-	0.6	8.9	3.6	-	3.6	-	12.5
4.6	-	2.1	-	-	1.3	-	-	-	0.6	9.1	3.6	1.7	5.3	-	14.4
-	-	0.2	-	-	-	-	-	-	-	0.2	0.3	4.1	4.4	-	4.6
21.0	-	32.3	1.5	0.3	4.0	1.4	0.5	-	17.0	107.0	3.9	9.0	12.9	-	119.9

Table 13. NET IMPORTS OF PETROLEUM PRODUCTS OECD AREA 1966-1970

Million tons

YEARS	L.P.G.	AVIATION FUEL	MOTOR GASOLINE	NAPHTHA	KEROSENE	GAS/ DIESEL OIL	FUEL OIL	OTHER PRODUCTS	TOTAL ALL PRODUCTS
EUROPEAN AREA									
1966	0.2	- 0.4	2.8	1.0[1]	0.1	14.5	11.6	1.0	30.8
1967	0.1	- 0.2	4.1	1.2	- 0.2	12.9	7.5	1.1	26.5
1968	-	- 0.2	2.3	0.6	- 0.6	12.3	5.3	0.9	20.6
1969	- 0.1	- 1.4	1.7	- 0.4	- 0.4	10.1	0.5	1.8	12.1
1970	- 0.1	- 2.0	1.9	- 0.9	- 0.4	12.9	1.5	0.9	13.8
NORTH AMERICA									
1966	- 0.7	4.6	2.0	4.4[1]	-	3.4	58.8	- 3.4	69.1
1967	- 0.9	3.9	2.1	- 0.2	0.2	3.9	61.4	- 4.3	66.1
1968	- 1.0	5.1	2.9	- 0.1	0.2	7.8	65.0	- 5.5	74.4
1969	- 1.3	5.9	3.2	0.1	0.3	9.0	71.1	6.5	94.8
1970	- 1.0	6.8	3.3	-	0.4	8.9	83.0	- 4.4	97.0
JAPAN									
1966	0.8	- 0.6	- 0.2	-[1]	- 0.2	2.3	7.5	0.5	11.5
1967	1.2	- 1.4	- 0.3	1.1	- 0.2	2.1	8.3	1.5	12.3
1968	1.7	- 1.7	- 0.1	2.3	- 0.2	2.4	9.0	1.9	15.3
1969	2.0	- 1.8	- 0.1	3.3	- 0.6	0.4	1.1	2.2	6.5
1970	2.7	- 1.1	- 0.1	6.5	- 0.1	1.5	5.6	2.9	17.9
AUSTRALIA									
1966	-	0.2	0.5	-	0.1	0.1	-	0.2	1.1
1967	-	0.1	0.3	0.1	0.1	0.1	0.1	0.1	0.9
1968	-	0.1	0.3	-	-	0.1	0.2	0.2	0.9
1969	-	0.1	0.4	0.1	0.1	0.2	1.2	0.2	2.3
1970	-	0.1	0.5	0.2	0.1	0.3	1.3	0.1	2.6

1. Light distillate for gasworks.

EUROPEAN AREA

Million tons

	1966	1967	1968	1969	1970
Crude oil, feedstocks and components					
Indigenous production	20.6	21.1	21.0	21.2	22.5
Imports	371.4	406.9	472.0	530.8	593.8
Total supply	392.2	428.0	493.0	552.0	616.3
Refinery intake	338.6	426.9	491.6	553.7	619.7
Exports	0.3	0.8	0.1	1.0	1.9
Total disposal	338.9	427.7	491.7	554.7	621.6
Petroleum products					
Refinery output	364.8	397.8	457.8	527.6	582.8
Imports	42.5	38.7	39.6	36.0	36.1
From other sources	2.5	1.4	1.5	1.6	1.6
Total supply	409.8	437.4	498.9	554.2	620.5
Inland consumption	354.9	385.1	434.1	486.6	644.0
Bunkers	29.9	30.9	33.4	38.1	37.4
Exports	7.6	9.4	14.9	18.9	18.3
Total disposal	392.4	425.4	482.4	543.6	699.7

Intra OECD European Area Trade

	1966	1967	1968	1969	1970
Crude oil, feedstocks, components	1.5	1.8	2.3	3.5	4.4
Petroleum products	67.7	0.8	75.0	84.4	97.9

NORTH AMERICA

	1966	1967	1968	1969	1970
Crude oil, feedstocks, components					
Indigenous production	495.9	480.9	499.8	509.8	536.4
Imports	70.8	63.4	70.9	70.1	62.2
Total supply	466.7	544.3	570.7	579.9	598.6
Refinery intake	539.9	562.0	596.4	615.5	633.8
Exports	18.2	24.0	23.0	27.5	32.7
Total disposal	558.1	538.0	619.4	643.0	676.5
Petroleum products					
Refinery output	534.5	558.1	589.2	606.5	626.4
Imports	71.2	76.4	84.4	93.2	109.0
From other sources	25.2	25.6	33.6	35.6	38.4
Total supply	630.9	660.1	707.2	735.3	773.8
Inland consumption	615.8	631.0	666.4	710.4	749.3
Bunkers	11.0	11.7	12.3	18.3	10.6
Exports	11.4	13.6	13.0	14.6	17.2
Total disposal	638.2	656.3	692.3	743.3	777.1

Intra North American Area Trade

	1966	1967	1968	1969	1970
Crude oil, feedstocks, components	17.1	20.8	22.9	27.4	33.0
Petroleum products	3.0	103.3	121.5	144.1	169.5

Table 14.   (Cont'd)

JAPAN

Million tons

	1966	1967	1968	1969	1970
Crude oil, feedstocks and components					
Indigenous production	0.7	0.7	0.7	0.7	0.8
Imports	97.7	103.3	121.5	144.1	169.5
Total supply	98.4	104.0	122.2	144.8	170.3
Refinery intake	84.2	102.6	115.6	137.6	169.2
Exports	1.7	–	–	–	–
Total disposal	85.9	102.6	115.6	137.6	169.2
Petroleum products					
Refinery output	82.1	101.1	114.1	139.8	159.7
Imports	13.2	14.8	17.7	18.6	27.4
From other sources	1.5	1.7	2.2	9.8	1.0
Total supply	96.8	117.6	134.0	168.2	188.1
Inland consumption	83.3	102.6	120.1	143.3	176.2
Bunkers	11.3	9.7	7.4	9.6	6.7
Exports	1.7	2.5	2.4	12.1	9.5
Total disposal	96.3	114.8	129.9	165.0	192.4

AUSTRALIA

Crude oil, feedstocks and components					
Indigenous production	0.4	1.0	1.8	2.1	8.1
Imports	17.9	19.5	20.0	20.2	15.9
Total supply	18.3	20.5	21.8	22.3	24.0
Refinery intake	18.2	20.3	22.0	22.2	23.8
Exports	–	–	–	–	0.1
Total disposal	18.2	20.3	22.0	22.2	23.9
Petroleum products					
Refinery output	16.1	17.9	19.6	19.7	21.1
Imports	1.1	0.9	0.9	2.3	2.6
From other sources	–	–	–	–	
Total supply	17.2	18.8	20.5	22.0	23.7
Inland consumption	14.3	15.5	17.1	19.0	19.9
Bunkers	1.8	2.1	2.3	2.2	2.3
Exports	1.1	1.1	1.1	0.8	1.4
Total disposal	17.2	18.7	20.5	22.0	23.6

Table 15. SUPPLY AND DISPOSAL OF CRUDE OIL, FEEDSTOCKS AND COMPONENTS AND PETROLEUM PRODUCTS

OECD AREA - 1970

Million tons

COUNTRIES	CRUDE OIL, FEEDSTOCKS AND COMPONENTS						PETROLEUM PRODUCTS							
	INDIGENOUS PRODUCTION	IMPORTS	TOTAL SUPPLY	REFINERY INTAKE	EXPORTS	TOTAL DISPOSAL	REFINERY OUTPUT	FROM OTHER SOURCES	IMPORTS	TOTAL SUPPLY	DELIVERIES TO INLAND CONSUMPTION	BUNKERS	EXPORTS	TOTAL DISPOSAL
	1	2	3 = 1 + 2	4	5	6 = 4+5	1	2	3	4 = 1 to 3	5	6	7	8
Austria	2.8	3.6	6.4	6.2	-	6.2	6.0	-	3.3	9.3	9.0	-	0.2	9.2
Belgium	-	29.9	29.9	29.9	-	29.9	28.2	-	6.1	34.3	22.3	2.7	9.1	34.1
Denmark	-	10.2	10.2	10.2	-	10.2	9.7	-	10.9	20.6	17.4	0.5	1.9	19.8
Finland	-	9.8	9.8	8.2	-	8.2	7.7	-	3.2	10.9	10.0	0.1	0.5	10.6
France	2.3	101.3	103.6	102.5	-	102.5	95.6	0.9	6.4	102.9	82.7	3.9	10.7	97.3
Germany	7.5	100.3	107.8	107.1	0.1	107.2	98.5	0.5	31.0	130.0	116.2	3.8	8.7	128.7
Greece	-	4.6	4.6	4.6	-	4.6	4.3	-	1.6	5.9	5.1	0.6	0.2	5.9
Iceland	-	-	-	-	-	-	-	-	0.5	0.5	0.5	-	-	0.5
Ireland	-	2.7	2.7	2.6	-	2.6	2.5	-	2.3	4.8	3.8	0.2	0.9	4.9
Italy	3.7	114.1	117.8	117.8	-	117.8	112.2	-	3.2	115.4	76.9	6.9	28.8	112.6
Luxembourg	-	-	-	-	-	-	-	-	1.4	1.4	1.4	-	-	1.4
Netherlands	1.9	60.1	62.1	62.2	0.6	62.8	58.5	-	9.6	68.1	24.0	8.8	33.8	66.6
Norway	-	6.5	6.5	5.8	-	5.8	5.6	-	4.2	9.8	7.4	0.6	1.6	9.6
Portugal	-	3.7	3.7	3.7	-	3.7	3.2	-	1.5	4.7	3.4	0.7	0.5	4.6
Spain	-	32.2	32.2	32.2	-	32.2	32.3	-	1.3	33.6	25.1	2.0	5.4	32.5
Sweden	-	11.8	11.8	11.8	-	11.8	11.4	-	20.6	32.0	28.2	1.2	1.8	31.2
Switzerland	-	5.5	5.5	5.5	-	5.5	5.2	-	7.8	13.0	12.3	-	0.3	12.6
Turkey	3.5	3.8	7.3	7.2	-	7.2	7.0	-	0.6	7.6	7.4	0.1	-	7.5
United Kingdom ...	0.1	102.1	102.2	102.0	1.2	103.2	94.7	0.2	20.1	115.0	90.8	5.5	17.4	113.7
Total OECD Europe .	21.8	602.2	624.0	619.5	1.9	621.4	582.6	1.6	135.6	719.8	543.9	37.6	121.8	703.3
Canada	62.0	29.1	91.1	62.9	32.0	94.9	60.9	2.8	10.1	73.9	67.7	2.3	3.7	73.7
United States	474.3	71.4	545.7	570.9	0.7	571.6	565.5	34.6	104.2	705.3	681.6	8.3	13.5	703.4
Total North America.	536.3	100.5	636.8	633.8	32.7	666.5	626.4	38.4	114.3	779.1	749.3	10.6	17.2	777.1
Japan	0.8	169.5	170.3	169.2	-	169.2	159.7	1.0	27.4	188.1	176.2	6.4	9.5	192.4
Australia	8.1	15.9	24.0	23.8	0.1	23.9	21.1	-	2.6	23.7	19.9	2.3	1.4	23.6
Total OECD Area ...	567.0	888.1	1,455.1	1,446.3	34.7	1,481.0	1,389.8	41.0	279.9	1,710.7	1,489.3	57.2	149.9	1,696.4

287

Table 16. OECD MEMBER COUNTRIES'
AND WORLD TANKER TONNAGE
1967 - 1971[1,2]

	MILLION DEADWEIGHT TONS					
	AT 1.1.67	AT 1.1.68	AT 1.1.69	AT 1.1.70	AT 1.1.71	AT 1.7.71
OECD Member countries						
United Kingdom ...	13.2	14.3	16.9	19.2	22.5	24.5
Norway	16.4	18.4	18.8	17.4	20.3	21.7
Japan	9.3	11.4	14.3	16.1	18.3	20.0
Greece	3.1	3.2	3.7	5.6	7.9	8.6
United States*	6.8	6.9	7.2	7.2	7.5	7.6
France	4.0	4.5	4.8	5.3	5.8	6.6
Italy	3.6	4.0	4.6	4.3	5.0	5.3
Sweden	3.1	3.3	3.6	3.6	3.6	3.8
Netherlands	2.7	3.1	3.2	3.1	3.5	3.3
Germany (F.R.)....	1.9	2.3	2.6	2.4	2.9	3.3
Spain	1.2	1.4	2.0	2.2	2.4	2.6
Denmark	1.9	2.0	2.4	2.5	2.3	2.3
Finland	0.5	0.6	0.8	0.9	1.1	1.1
Others OECD	1.2	1.4	1.4	1.3	1.6	1.8
Total OECD	68.9	76.8	86.3	91.1	104.7	112.5
Others outside OECD						
Liberia	23.6	27.0	31.3	35.3	42.1	45.2
Panama	4.0	4.5	5.0	5.3	5.5	6.5
USSR	3.9	4.3	4.6	4.1	4.4	4.4
Rest of the World ..	4.6	5.0	6.3	7.3	8.0	7.4
Total World	105.0	117.6	133.5	143.1	164.7	176.0

*   Commercial Fleet only.
1.   Vessels of over 7,000 tons deadweight.
2.   Includes combination carriers.
SOURCE: Davies and Newman.

Table 17
(See pp. 290 and 291)

Table 17.   CRUDE OIL EXPORTED TO THE OEC

EXPORTING COUNTRIES	AUSTRIA	BELGIUM	DENMARK	FINLAND	FRANCE	GERMANY	GREECE	ICELAND	IRELAND
Venezuela	-	2.8	0.4	0.1	2.4	3.4	-	-	
Netherlands Antilles	-	-	-	-	-	-	-	-	-
Other Western Hemisphere	-	-	-	-	-	-	-	-	-
Libya	0.4	6.1	1.3	-	17.6	40.9	-	-	-
Algéria	-	1.5	0.1	-	27.0	8.0	-	-	-
Nigeria	-	0.7	0.8	-	5.2	6.9	-	-	-
Other Africa	0.1	0.2	1.0	-	2.1	1.0	-	-	-
Kuwait	-	5.2	2.8	-	11.1	3.9	-	-	0.9
Saudi Arabia	0.1	5.1	1.4	-	9.4	12.1	-	-	0.8
Iran	-	3.5	0.7	2.8	3.8	8.3	-	-	0.7
Iraq	0.5	1.1	-	-	12.2	3.5	-	-	0.2
Qatar	-	1.1	-	-	1.8	0.3	-	-	-
Other near and Middle East	0.1	1.8	1.5	-	7.3	7.0	-	-	-
Far East	-	0.1	-	-	-	-	-	-	-
TOTAL	1.2	29.2	10.0	2.8	99.9	95.3	-	-	2.6

IMPOR TI

290

Million tons

OUNTRIES

ITALY	LUXEMBOURG	NETHERLANDS	NORWAY	PORTUGAL	SPAIN	SWEDEN	SWITZERLAND	TURKEY	UNITED KINGDOM	EUROPE TOTAL	CANADA	UNITED STATES	NORTH AMERICA TOTAL	JAPAN	OECD TOTAL
2.2	-	1.2	1.2	-	2.1	1.3	-	-	4.6	21.7	18.6	14.0	32.6	0.6	54.3
-	-	-	-	-	-	-	-	-	0.2	0.2	-	-	-	-	0.2
-	-	-	-	-	-	-	-	-	0.3	0.3	1.0	1.1	2.1	-	2.4
35.6	-	12.2	0.6	-	8.0	0.4	3.2	0.5	23.7	150.5	-	2.3	2.3	0.3	153.1
2.8	-	0.2	0.1	-	1.2	-	0.7	-	1.3	42.9	-	0.3	0.3	-	43.2
0.5	-	7.7	1.1	-	2.2	3.1	-	-	7.6	35.8	2.5	2.4	4.9	-	40.7
-	-	1.3	-	0.1	0.2	-	0.2	-	0.5	6.7	-	-	-	0.6	7.3
13.8	-	11.6	0.7	-	2.1	0.7	0.6	-	26.1	79.5	7.0	1.7	2.4	23.2	105.1
16.3	-	9.2	1.1	0.6	8.5	-	0.6	0.4	16.1	81.7	1.9	2.1	4.0	33.3	119.0
6.4	-	6.1	0.2	0.1	2.3	1.1	-	-	8.4	44.4	2.6	1.7	4.3	73.7	122.4
21.6	-	5.0	-	2.0	1.4	0.2	-	2.9	2.4	53.0	1.2	-	1.2	-	54.2
1.4	-	1.2	0.1	-	-	0.8	-	-	4.2	10.9	-	-	-	0.2	11.1
4.5	-	2.6	1.4	0.8	4.2	3.1	-	0.1	4.3	38.7	0.7	4.2	4.9	14.3	57.9
-	-	-	-	-	-	-	-	-	-	0.1	-	3.5	3.5	22.0	25.6
105.1	-	58.3	6.5	3.6	32.2	10.7	5.3	3.9	99.7	566.4	29.2	33.3	62.5	168.2	797.1

Table 18. FOB VALUE OF OIL EXPORTS FROM DEVELOPING COUNTRIES
AND OIL EXPORTS AS A PERCENTAGE OF TOTAL EXPORTS – 1970

Millions of dollars

DEVELOPING AREAS[2]	VALUE OF OIL[1] EXPORTED WORLD-WIDE	TOTAL WORLD EXPORTATION ALL COMMODITIES	OIL AS % OF TOTAL EXPORTS	VALUE OF OIL EXPORTED TO OECD COUNTRIES	TOTAL EXPORTS TO OECD COUNTRIES	OIL EXPORTS AS % OF TOTAL EXPORTS TO OECD COUNTRIES
Africa[3]	4,060	12,310	32.9	3,675	9,967	36.8
Latin America[4] ...	4,310	17,590	24.5	2,690	12,710	21.1
Middle East[5] ...	8,730	10,240	85.2	6,604	7,378	89.5
Others[6]	960	14,150	6.7	431	8,805	4.8
Total	18,060	54,290	33.2	13,400	38,860	34.4

1. May include small quantities of fuel other than oil.
2. Sum of regions other than: North America, Western Europe, Australia, New Zealand, South Africa and Japan, Eastern Europe, China, Mongolia, North Korea and North Vietnam.
3. Excluding South Africa.
4. Excluding Caribbean Islands.
5. UN definition.
6. Obtained by subtracting Africa, Latin America and Middle East from Total.
SOURCE: UN Monthly Bulletin of Statistics, July 1972, Special Table C.

Table 19. CIF[1] VALUE OF OIL IMPORTED BY OECD COUNTRIES[2] FROM COUNTRIES IN PROCESS OF DEVELOPMENT AND OIL IMPORTS AS A PERCENTAGE OF TOTAL IMPORTS - 1970

Million dollars

OECD MEMBER COUNTRIES	OIL IMPORTS (CRUDE, PARTLY REFINED AND PRODUCTS)	TOTAL IMPORTS (ALL COMMODITIES)	OIL IMPORTS AS % OF TOTAL IMPORTS
Austria	26	3, 549	0.7
Belgium/Luxembourg ..	520	11, 362	4.5
Canada	516	13, 348	3.8
Denmark	198	4, 385	4.5
Finland	52	2, 637	1.9
France	1, 662	18, 923	8.7
Germany	1, 598	29, 814	5.3
Greece	69	1, 958	3.5
Iceland	1	157	0.6
Ireland	53	1, 569	3.3
Italy	1, 672	14, 939	11.1
Japan	2, 632	18, 881	13.9
Netherlands	1, 168	13, 393	8.7
Norway	15[3]	3, 698	n.a.
Portugal	99	1, 590	6.2
Spain	504	4, 715	10.6
Sweden	234	7, 005	3.3
Switzerland	99	6, 448	1.5
Turkey	50	891	5.6
United Kingdom	1, 733	21, 678	7.9
United States	1, 909	39, 952	4.7
Total	14, 810	220, 892	6.7

1. With the exception of United States and Canada for which FOB values are shown.
2. Excluding Australia.
3. Oil products only.
SOURCE: OECD Trade by Commodities - Series C - January-December, 1970.

OECD SALES AGENTS
DEPOSITAIRES DES PUBLICATIONS DE L'OCDE

ARGENTINE
Libreria de las Naciones
Alsina 500, BUENOS AIRES.

AUSTRALIA – AUSTRALIE
B.C.N. Agencies Pty, Ltd.,
178 Collins Street, MELBOURNE 3000.

AUSTRIA – AUTRICHE
Gerold and Co., Graben 31, WIEN 1.
Sub-Agent: GRAZ: Buchhandlung Jos. A. Kien-
reich, Sackstrasse 6.

BELGIUM – BELGIQUE
Librairie des Sciences
Coudenberg 76-78 et rue des Eperonniers 56,
B 1000 BRUXELLES 1.

BRAZIL – BRESIL
Mestre Jou S.A., Rua Guaipá 518,
Caixa Postal 24090, 05000 SAO PAULO 10.
Rua Senador Dantas 19 s/205-6, RIO DE
JANEIRO GB.

CANADA
Information Canada
OTTAWA.

DENMARK – DANEMARK
Munksgaard International Booksellers
Nørregade 6, DK-1165 COPENHAGEN K

FINLAND – FINLANDE
Akateeminen Kirjakauppa, Keskuskatu 2,
HELSINKI.

FORMOSA – FORMOSE
Books and Scientific Supplies Services, Ltd.
P.O.B. 83, TAIPEI,
TAIWAN.

FRANCE
Bureau des Publications de l'OCDE
2 rue André-Pascal, 75775 PARIS CEDEX 16
Principaux sous dépositaires :
PARIS : Presses Universitaires de France,
49 bd Saint-Michel, 75005 Paris.
Sciences Politiques (Lib.)
30 rue Saint-Guillaume, 75007 Paris.
13100 AIX-EN-PROVENCE : Librairie de l'Uni-
versité.
38000 GRENOBLE : Arthaud.
67000 STRASBOURG : Berger-Levrault.
31000 TOULOUSE : Privat.

GERMANY – ALLEMAGNE
Deutscher Bundes-Verlag G.m.b.H.
Postfach 9380, 53 BONN.
Sub-Agents: BERLIN 62: Elwert & Meurer.
HAMBURG : Reuter-Klöckner ; und in den
massgebenden Buchhandlungen Deutschlands.

GREECE – GRECE
Librairie Kauffmann, 28 rue du Stade,
ATHENES 132.
Librairie Internationale Jean Mihalopoulos et Fils
75 rue Hermou, B.P. 73, THESSALONIKI.

ICELAND – ISLANDE
Snæbjörn Jónsson and Co., h.f., Hafnarstræti 9,
P.O.B. 1131, REYKJAVIK.

INDIA – INDE
Oxford Book and Stationery Co. :
NEW DELHI, Scindia House.
CALCUTTA, 17 Park Street.

IRELAND – IRLANDE
Eason and Son, 40 Lower O'Connell Street,
P.O.B. 42, DUBLIN 1.

ISRAEL
Emanuel Brown :
9, Shlomzion Hamalka Street, JERUSALEM.
35 Allenby Road, and 48 Nahlath Benjamin St.,
TEL-AVIV.

ITALY – ITALIE
Libreria Commissionaria Sansoni :
Via Lamarmora 45, 50121 FIRENZE.
Via Bartolini 29, 20155 MILANO.
sous-dépositaires :
Editrice e Libreria Herder,
Piazza Montecitorio 120, 00186 ROMA.
Libreria Hoepli, Via Hoepli 5, 20121 MILANO.
Libreria Lattes, Via Garibaldi 3, 10122 TORINO.
La diffusione delle edizioni OCDE è inoltre assicu-
rata dalle migliori librerie nelle città più importanti.

JAPAN – JAPON
Maruzen Company Ltd.,
6 Tori-Nichome Nihonbashi, TOKYO 103,
P.O.B. 5050, Tokyo International 100-31.

LEBANON – LIBAN
Redico
Immeuble Edison, Rue Bliss, B.P. 5641
BEYROUTH.

THE NETHERLANDS – PAYS-BAS
W.P. Van Stockum
Buitenhof 36, DEN HAAG.

NEW ZEALAND – NOUVELLE-ZELANDE
Government Printing Office,
Mulgrave Street (Private Bag), WELLINGTON
and Government Bookshops at
AUCKLAND (P.O.B. 5344)
CHRISTCHURCH (P.O.B. 1721)
HAMILTON (P.O.B. 857)
DUNEDIN (P.O.B. 1104).

NORWAY – NORVEGE
Johan Grundt Tanums Bokhandel,
Karl Johansgate 41/43, OSLO 1.

PAKISTAN
Mirza Book Agency, 65 Shahrah Quaid-E-Azam,
LAHORE 3.

PORTUGAL
Livraria Portugal, Rua do Carmo 70, LISBOA.

SPAIN – ESPAGNE
Mundi Prensa, Castelló 37, MADRID 1.
Libreria Bastinos de José Bosch, Pelayo 52,
BARCELONA 1.

SWEDEN – SUEDE
Fritzes, Kungl. Hovbokhandel,
Fredsgatan 2, 11152 STOCKHOLM 16.

SWITZERLAND – SUISSE
Librairie Payot, 6 rue Grenus, 1211 GENEVE 11
et à LAUSANNE, NEUCHATEL, VEVEY,
MONTREUX, BERNE, BALE, ZURICH.

TURKEY – TURQUIE
Librairie Hachette, 469 Istiklal Caddesi, Beyoglu,
ISTANBUL et 12 Ziya Gökalp Caddesi, ANKARA.

UNITED KINGDOM – ROYAUME-UNI
H.M. Stationery Office, P.O.B. 569, LONDON
SE1 9NH
or
49 High Holborn
LONDON WC1V 6HB (personal callers)
Branches at: EDINBURGH, BIRMINGHAM,
BRISTOL, MANCHESTER, CARDIFF,
BELFAST.

UNITED STATES OF AMERICA
OECD Publications Center, Suite 1207,
1750 Pennsylvania Ave, N.W.
WASHINGTON, D.C. 20006. Tel. : (202)298-8755.

VENEZUELA
Libreria del Este, Avda. F. Miranda 52,
Edificio Galipan, CARACAS.

YUGOSLAVIA – YOUGOSLAVIE
Jugoslovenska Knjiga, Terazije 27, P.O.B. 36,
BEOGRAD.

Les commandes provenant de pays où l'OCDE n'a pas encore désigné de dépositaire
peuvent être adressées à :
OCDE, Bureau des Publications, 2 rue André-Pascal, 75775 Paris CEDEX 16
Orders and inquiries from countries where sales agents have not yet been appointed may be sent to
OECD, Publications Office, 2 rue André-Pascal, 75775 Paris CEDEX 16

OECD PUBLICATIONS
2, rue André-Pascal
75775 PARIS CEDEX 16
No. 31,515.   1973.

PRINTED IN FRANCE